GW00630704

Aviation Law
for Pilots

Aviation Law for Pilots

Tenth Edition

R.B. Underdown and Tony Palmer

**Blackwell
Science**

© the estates of S.E.T. Taylor & H.A. Palmer 1971,
    1974, 1978, 1983
© the estates of S.E.T. Taylor, H.A. Palmer &
    R.B. Underdown 1986, 1988, 1989, 1990
Eighth edition © the estate of R.B. Underdown
    1993
Ninth edition © the estate of R.B. Underdown &
    T. Palmer 1995
Ninth edition revised © the estate of
    R.B. Underdown & T. Palmer 1997
Tenth edition © Blackwell Science Ltd

Blackwell Science Ltd
Editorial Offices:
Osney Mead, Oxford OX2 0EL
25 John Street, London WC1N 2BL
23 Ainslie Place, Edinburgh EH3 6AJ
350 Main Street, Malden
    MA 02148 5018, USA
54 University Street, Carlton
    Victoria 3053, Australia
10, rue Casimir Delavigne
    75006 Paris, France

Other Editorial Offices:

Blackwell Wissenschafts-Verlag GmbH
Kurfürstendamm 57
10707 Berlin, Germany

Blackwell Science KK
MG Kodenmacho Building
7-10 Kodenmacho Nihombashi
Chuo-ku, Tokyo 104, Japan

The right of the Author to be identified as the
Author of this Work has been asserted in
accordance with the Copyright, Designs and
Patents Act 1988.

All rights reserved. No part of this publication
may be reproduced, stored in a retrieval system,
or transmitted, in any form or by any means,
electronic, mechanical, photocopying, recording
or otherwise, except as permitted by the UK
Copyright, Designs and Patents Act 1988, without
the prior permission of the publisher.

First, second and third editions published in Great
Britain by Crosby Lockwood & Son Ltd, 1971–78
Fourth edition published by Granada Technical
Books 1983
Fifth edition published by Collins Professional
and Technical Books 1986
Sixth, seventh, eighth, ninth and ninth (revised)
editions published by Blackwell Science, 1988–97
Tenth edition published 2000

Set in Palatino and produced by Gray Publishing,
Tunbridge Wells, Kent
Printed and bound in Great Britain by
MPG Books Ltd, Bodmin, Cornwall

The Blackwell Science logo is a
trade mark of Blackwell Science Ltd,
registered at the United Kingdom
Trade Marks Registry

DISTRIBUTORS

    Marston Book Services Ltd
    PO Box 269
    Abingdon
    Oxon OX14 4YN
    (*Orders*: Tel: 01235 465500
                Fax: 01235 465555)

USA
    Blackwell Science, Inc.
    Commerce Place
    350 Main Street
    Malden, MA 02148-5018
    (*Orders*: Tel: 800 759 6102
                     781 388 8250
                Fax: 781 388 8255)

Canada
    Login Brothers Book Company
    324 Saulteaux Crescent
    Winnipeg, Manitoba R3J 3T2
    (*Orders*: Tel: 204 837-2987
                Fax: 204 837-3116)

Australia
    Blackwell Science Pty Ltd
    54 University Street
    Carlton, Victoria 3053
    (*Orders*: Tel: 03 9347 0300
                Fax: 03 9347 5001)

A catalogue record for this title is available from
the British Library

ISBN 0-632-05335-6

Library of Congress
Cataloging-in-Publication Data is available

For further information on
Blackwell Science, visit our website:
www.blackwell-science.com

Contents

Preface

This is a completely new book. It follows the general chapter arrangement of the previous nine editions, but every chapter has been completely rewritten. The objective has been to take account of the Joint Airworthiness Requirements (JAR) as they affect aviation law at this moment. These are still by no means complete. There are number of areas that are still to be agreed between the numerous member nations of the Joint Aviation Authorities (JAA) and this can be a very long and tedious process. There are further problems ahead because the JAA is due to be replaced by the European Aviation Safety Authority (EASA) some time in the next five years. At the moment there is no legal or constitutional precedent for it. Effectively it will be the first European federal executive body devolved from the European Commission. Many JAA nations are still finding it difficult to implement JARs through their own legal systems and so they will find it even harder to absorb legally, politically and constitutionally the loss of sovereignty which the creation of the EASA implies.

The aim has been to produce a reference source for everyone concerned with the flying or operating of civil aircraft at the beginning of the twenty-first century.

Tony Palmer
Southampton

Abbreviations

AAIB	Air Accidents Investigation Branch
AAL or aal	Above Aerodrome Level
ABN	Aerodrome Beacon (light)
A/C or a/c	Aircraft
ACAS	Airborne Collision Avoidance System
ACC	Area Control Centre
AD	Aerodrome
ADA	Advisory Airspace
ADF	Automatic Direction Finder
ADR	Advisory Airspace
ADT	Approved Departure Time
AFI	Assistant Flying Instructor
AFM	Affirm
AFIS	Aerodrome Flight Information Service
AFS	Aeronautical Fixed Service
AFTN	Aeronautical Fixed Telecommunication Network
A/G	Air to Ground
agl	Above Ground Level
AGL	Aeronautical Ground Light
AIAA	Area of Intense Air Activity
AIC	Aeronautical Information Circular
AIP	Aeronautical Information Publication
AIRAC	Aeronautical Information Regulation and Control
AIREP	Air Report
AIS	Aeronautical Information Service
ALERFA	Alert Phase (Search and Rescue)
ALS	Approach Lighting System
AMC	Airspace Management Cell
AMC	Acceptable Means of Compliance (JAR)
AME	Authorised Medical Examiner
AMO	Approved Maintenance Organisation
AMS	Aeronautical Mobile Service
amsl	Above Mean Sea Level
AN(G)R	Air Navigation (General) Regulations
ANO	Air Navigation Order
AO	Aircraft Operator
AOC	Air Operator's Certificate

AOM	Aerodrome Operating Minima
ARP	Aerodrome Reference Point
ARN	ATS Route Network
ASDA	Accelerate Stop Distance Available
ASR	Altimeter Setting Region
ATA	Actual Time of Arrival
ATC	Air Traffic Control
ATCC	Air Traffic Control Centre
ATCU	Air Traffic Control Unit
ATD	Actual Time of Departure
ATFM	Air Traffic Flow Management
ATIS	Automatic Terminal Information Service
ATS	Air Traffic Service
ATSU	Air Traffic Service Unit
ATZ	Aerodrome Traffic Zone
AUS	Airspace Utilisation Service
Authority	Civil Aviation Authority (in the UK)
AWD	Airworthiness Division (of the CAA)
AVGAS	Aviation Gasoline (Petrol)
AVTUR	Aviation Turbine Fuel (kerosene)
AWY	Airway
B-RNAV	Basic Area Navigation Radio System
C	Centre (runway identification)
C	Degrees Celsius
CAA	Civil Aviation Authority
CANP	Civil Aircraft Notification Procedure
CAP	Civil Aviation Publication
CAS	Controlled Airspace
CAT	Clear Air Turbulence
CAVOK	Cloud, Visibility and Weather OK
CDL	Configuration Deviation List
CDR	Conditional Route
CEU	Central Executive Unit (Flow Management)
C of A	Certificate of Airworthiness
COSPAS	Cosmos Rescue System (Russian)
CRAM	CDR Availability Message
CTA	Control Area
CTMO	Central Traffic Management Organisation
CTOT	Calculated Take-off Time
CTR	Control Zone
D	Danger Area (followed by identification)
DA	Decision Altitude (Precision Approaches)
DAAIS	Danger Area Activity Information Service

DACS	Danger Area Crossing Service
DBE	Eurocontrol Data Bank
DF	Direction Finding
DH	Decision Height (Precision Approaches}
DME	Distance Measuring Equipment
DOC	Designated Operational Coverage (radio aids)
EASA	European Aviation Safety Authority (to replace JAA)
EAT	Expected Approach Time
EC	European Community
ECA	Emergency Controlling Authority
ECAC	European Civil Aviation Conference
EDA	Emergency Distance Available (same as ASDA)
EET	Estimated Elapsed Time
EFIS	Electronic Flight Instrument System
ELBA	Emergency Location Beacon (aircraft)
ELT	Emergency Locator Transmitter (aircraft)
EOBT	Estimated Off-Block Time
EPIRB	Emergency Position Indicating Radio Beacon
ETA	Estimated Time of Arrival
ETOPS	Extended Range Twin-jet Operations
FAA	Federal Aviation Administration (USA)
FAF	Final Approach Fix
FAR	Federal Aviation Regulations (USA)
FAX	Facsimile Transmission
FCL	Flight Crew Licensing
FDO	Flight Data Operations (Air traffic flow)
FI	Flying Instructor
FIC	Flight Information Centre
FIR	Flight Information Region
FIS	Flight Information Service
FL	Flight Level
FLPFM	Foot Launched Powered Flying Machine
FMP	Flow Management Position (Air Traffic Flow)
FMU	Flow Management Unit
FPL	Filed Flight Plan (message designator)
FTL	Flight Time Limitations
GAT	General Air Traffic
GCA	Ground Controlled Approach (System)
GMC	Ground Movement Control
GMT	Greenwich Mean Time (UTC)
GNSS	Global Navigation Satellite System
GP	Glide Path (ILS)
GPS	Geographical Positioning System (Satellite)

H24	Continuous Day and Night Service
HDG	Heading
HF	High Frequency (3–30 MHz)
HI	High Intensity Directional Lights (Runway approach)
HIRL	High Intensity Runway Lighting
HIRTA	High Intensity Radio Transmission Area
HJ	Sunrise to Sunset (Heures Jour)
HN	Sunset to Sunrise (Heures Nuit)
HO	Service available to meet Operational requirements
HR	Hours
HX	No specific working hours
Hz	Hertz (cycles per second)
IAC	Instrument Approach Chart
IAF	Initial Approach Fix
IAS	Indicated Air Speed
IASTA	International Air Services Transport Agreement
IATA	International Air Transport Agreement
IATA	International Air Transport Association
IBN	Identification Beacon (light)
ICAO	International Civil Aviation Organisation
ID	Identification Document
IEM	Interpretive/Explanatory Material (JAR)
IFPS	Integrated Flight Planning System
IFPU	Integrated Flight Planning Unit
IFR	Instrument Flight Rules
ILS	Instrument Landing System
IM	Inner Marker (ILS)
IMC	Instrument Meteorological Conditions
INCERFA	Uncertainty Phase (Search and Rescue)
INS	Inertial Navigation System
IR	Instrument Rating
IRE	Instrument Rating Examiner
IRVR	Instrumented Runway Visual Range
ISA	International Standard Atmosphere
JAA	Joint Aviation Authorities
JAR	Joint Aviation Requirements
kg	Kilograms
kHz	Kilohertz (preceded by number)
km	Kilometre
KT or kt	Knots (preceded by figure)
L	Left (runway identification)
L	Locator (low powered NDB)
LARS	Lower Airspace Radar Advisory Service

LDA	Landing Distance Available
LF	Low Frequency (30–300 kHz)
LITAS	Low Intensity Two Colour Approach System
LLZ	Localiser (ILS)
LMC	Last Minute Changes (Load Sheets)
LOFT	Line Orientated Flight Training
LRNS	Long Range Navigation System
LVP	Low Visibility Procedures
M	Mach Number (followed by figure)
m	metres (preceded by figure)
MAA	Maximum Authorised Altitude (Airways)
MAP	Missed Approach Point
MATZ	Military Aerodrome Traffic Zone
mb	Millibars (preceded by figure)
MDA	Minimum Descent Altitude (Non-precision Approaches)
MDH	Minimum Descent Height (Non-precision Approaches)
MEA	Minimum En-route Altitude (Airways)
MEDA	Military Emergency Diversion Aerodrome
MEHT	Minimum Eye Height over Threshold (for VASI and PAPI)
MEL	Minimum Equipment List
METAR	Aviation Routine Weather Report (in international code)
MF	Medium Frequency (300–3000 kHz)
MHz	Megahertz (preceded by number)
min	Minutes
MKR	Marker (radio beacon)
MM	Middle Marker (ILS)
MNPS	Minimum Navigation Performance Specification
MOTNE	Meteorological Operational Telecommunication Network Europe
MSLM	Maximum Structural Landing Mass
MSTOM	Maximum Structural Take-off Mass
MZFM	Maximum Zero Fuel Mass
Ms	Minus
MSA	Minimum Sector Altitude
MWO	Meteorological Watch Office
NAP	Noise Abatement Procedure
NATS	National Air Traffic Service
NATZ	Notified Aerodrome Traffic Zone
NDB	Non-directional Beacon
NM or nm	Nautical Mile (preceded by number)
NMC	National Meteorological Centre
NOH	Notified Operating Hours
NOSIG	No Significant Change (Trend forecasts)

NOTAM	Notice to Airmen
NPA	Notice of Proposed Amendment (JAR)
NSF	Non-scheduled Flights
OAC	Oceanic Area Control Centre
OAT	Operational Air Traffic
OCA	Oceanic Control Area
OCA (H)	Obstacle Clearance Altitude (Height)
OCL	Obstacle Clearance Limit
OM	Outer Marker (ILS)
OM	Operations Manual
P....	Prohibited Area (followed by identification)
PANS	Procedures for Air Navigation Services (ICAO)
PAPI	Precision Approach Path Indicator (Lighting system)
PAR	Precision Approach Radar
PAX	Passengers
PIC	Pilot In Charge
POB	Persons On Board
PPR	Prior Permission Required (Aerodromes)
Ps	Plus
PT	Public Transport
QDM	Magnetic Heading to reach station (in zero wind)
QDR	Magnetic Bearing from station
QFE	Pressure at aerodrome elevation or at runway threshold
QNH	Altimeter setting to get aerodrome elevation when on ground
QR	Quadrantal Rules (for altitude separation)
QTE	True Bearing from satation
R	Restricted Area (followed by identification)
R	Right (Runway)
RA	Resolution Advisory (ACAS)
RAS	Radar Advisory Service
RCC	Rescue Co-ordination Centre
RIS	Radar Information Service
RLCE	Request Level Change En-route
RLLC	Royal Low Level Corridor
RNAV	Area Navigation Radio System
RNOTAM	Royal NOTAM
RPL	Repetitive Flight Plan
RTF	Radio Telephony
RTHL	Runway Threshold Lights
RTR	Radar Termination Range
RTZL	Runway Touchdown Zone Lights
RVR	Runway Visual Range

RVSM	Reduced Vertical Separation Minima
RWY	Runway
SAR	Search and Rescue
SARPS	Standards and Recommended Practices (ICAO)
SARSAT	Search and Rescue Satellite Aided Tracking System
SCR	Semi-circular Rules (altitude separation)
sec	Seconds
SELCAL	Selective Calling System
SFC	Surface
SHF	Super High Frequency (3000–30000 MHz)
SID	Standard Instrument Departure
SIGMET	Message regarding en-route weather likely to affect aircraft safety
SITA	Societé Internationale de Telecommunications Aeronautique
SLMG	Self Launched Motor Glider
SMC	Surface Movement Control
SNOCLO	Closed by Snow (in VOLMET message)
SNOWTAM	Special NOTAM dealing with snow on aerodromes
SMR	Surface Movement Radar
SPECI	Aviation Selected Special Weather Report (in code)
SRA	Surveillance Radar Approach
SRR	Search and Rescue Region
SSR	Secondary Surveillance Radar
STAR	Standard Terminal Arrival Route
STOL	Short Take-off and Landing
SVFR	Special Visual Flight Rules
TA	Transition Altitude
TA	Traffic Advisory (ACAS)
TACAN	UHF Tactical Air Navigation Aid (Military)
TAF	Aerodrome Forecast
TAS	True Airspeed
TBC	Tactical Booking Cell
TCAS	Airborne Collision Avoidance System
TDA	Temporary Danger Area
TDZ	Touch Down Zone
TEMPO	Temporarily (Meteorology trend message)
TMA	Terminal Control Area
TMG	Touring Motor Glider (JAR)
TOC	Top Of Climb
TOD	Top Of Descent
TODA	Take-off Distance Available
TORA	Take-off Run Available

TR	Type Rating
TRA	Temporary Restricted Area
TRE	Type Rating Examiner
TRL	Transition Level
TVOR	Terminal VOR
TWIL	Twilight (civil)
TWR	Tower (aerodrome control)
TWY	Taxiway
UAR	Upper Air Route
UHF	Ultra Frequency (300–3000 MHz)
UIR	Upper Information Region
UK	United Kingdom
U/S	Unserviceable
UTA	Upper Control Area
UTC	Co-ordinated Universal Time (GMT)
VAR	Magnetic Variation
VASIS	Visual Approach Slope Indicator System
VDF	VHF DF system
VFR	Visual Flight Rules
VHF	Very High Frequency (30–300 MHz)
VIS	Visibility
VLF	Very Low Frequency (3–30 kHz)
VLR	Very Long Range
VMC	Visual Meteorological Conditions
VOLMET	Meteorological Broadcasts for Aircraft in Flight
VOR	VHF Omni-directional Range
VORTAC	VOR combined with TACAN
VOT	Test facility for aircraft VOR equipment
VSTOL	Very Short Take-off and Landing
VTOL	Vertical Take-off and Landing
WEF	With Effect From
WIP	Work In Progress
WPT	Waypoint
Z	UTC (in meteorological message)

Chapter 1
International and UK Air Law

Introduction

Ever since Bleriot flew from France to England at the beginning of the twentieth century, it has been obvious that air law has to be an international business. In aviation no country can divorce itself from the world outside. After the World War I an international convention agreed a set of common aviation laws. The United States was a notable absentee from this meeting. This was of little importance, as between the wars there was hardly any transatlantic air traffic. After World War II, a meeting in Chicago laid the foundations of what we now know as the International Civil Aviation Organisation (ICAO).

Controlling Influences on UK Air Law

The main international authorities that affect UK air law are:

- *ICAO (International Civil Air Organisation)*
  with representatives from practically every state involved in international aviation.
- *JAA (Joint Aviation Authorities)*
  representing all of the European Union (EU) plus a number of other countries in and around Europe such as Iceland, Switzerland and Turkey. Around about 2005, it is proposed that JAA will be replaced by the European Aviation Safety Authority (EASA). This will be a federal executive body derived from the EU. This is only a proposal and no doubt there will be many hurdles to be crossed before it becomes effective.
- *EC (European Union)*
  at present 15 nations enact laws that affect all members. It is expected that they will gradually become involved with aviation law, as noted in the above paragraph.

ICAO

The main convention of ICAO has 18 annexes which lay down:

- *Standards*
  which all member states are expected to incorporate in their aviation law.
- *Recommended practices*
  the name is self-explanatory.

Together these are known as SARPs. If countries cannot comply with them, ICAO should be notified and this is often done in a country's Aeronautical Information Publication (AIP). Notes will also be found in flight guides (AERADs and Jeppesen) giving where the Air Traffic Control (ATC) regulations differ from ICAO SARPs.

Joint Aviation Requirements (JAR)

These will eventually cover all the aviation rules applicable to all the participating states. Areas of aviation law that are not covered by JAR will still be subject to the relevant National Rules. For example, in 1998 Flight Time Limitations (FTL) are still noted as 'reserved' in JAR-OPS. So in the UK and other JAA countries the existing rules will be complied with. Before new regulations become law an NPA (Notice of Proposed Amendment) will be issued and time allowed for discussion and possible variations of the proposed regulations to be made.

International Air Transport Agreement (IATA)

This is not to be confused with the International Air Transport Association (IATA) which is an association of the world's principal airlines. Under the auspices of ICAO, in 1948, Five Freedoms of the Air were suggested. Considering states A, B and C these are:

- *First freedom*
  the right of aircraft from state A to fly over state B without landing.
- *Second freedom*
  the right of aircraft from state A to land in state B for technical reasons.

These two freedoms, sometimes called technical rights, are fairly widely accepted, but not by every ICAO member. The full name for them is the International Air Services Transport Agreement (IASTA).

- *Third freedom*
  the right of aircraft from state A to accept paying traffic from A and put it down in state B.
- *Fourth freedom*
  the right of aircraft from state A to pick up paying traffic in state B and put it down in its own state (A).

These two freedoms obviously go hand in hand and are the subject of many bilateral agreements between states. When it is realised that there are about 190 members of ICAO, the number of bilaterals that would be

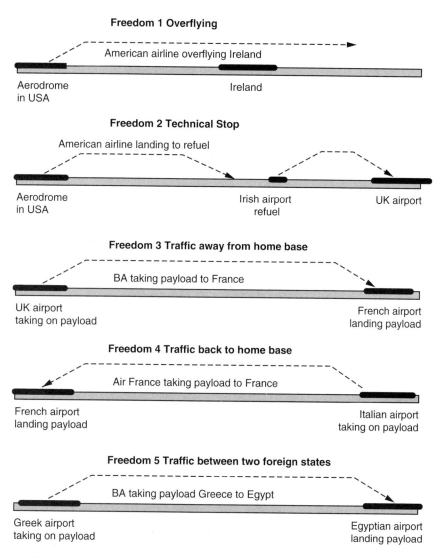

Figure 1.1 The five freedoms.

required if every state wanted to have freedom to operate commercially to every other state can easily be visualised.

- *Fifth freedom*
  the right of aircraft from state A to take paying traffic from state B to state C.

When the IATA was first being discussed, only very few nations were in favour of the fifth freedom, but gradually, in some special cases, attempts have been made to implement it. Details of which freedoms are accepted by particular states will be found in Jeppesen Airways manuals and Civil Aviation Publication (CAP) 555 published by the CAA and titled 'Overseas Flight Clearance Guide'. AIPs for each country will give details of the freedoms affecting that country. Figure 1.1 gives some practical examples of the five freedoms.

Tokyo Convention (1963)

Under ICAO auspices, this agreed on rules governing the action to be taken in the event of a hijacking.

Relevant JAR (Joint Aviation Requirements)

- *JAR 23 and 25*
  these deal with smaller and large aircraft, respectively. Aircraft certificated under these will be acceptable to all the JAA states. Recently, the JAA and the American Federal Aviation Administration (FAA) have been working on 'Cooperative and Concurrent Certification' so as to be able to end up with an aircraft acceptable to both authorities. The equivalent American regulations are FAR 23 and 25.
- *JAR 145*
  this deals with Approved Maintenance Organisations (AMO). These organisations will be able to issue Certificates of Airworthiness (C of A) acceptable to all JAA states.
- *JAR-FCL 1 and 2*
  these are for Flight Crew Licensing for Aeroplanes and Helicopters, respectively. They deal in great detail with examinations, privileges, medical requirements, etc. The full requirements are gradually being introduced into the UK. It is proposed that they will come fully into effect by July 1999 for aeroplanes and January 2000 for helicopters. Eventually licences issued under these requirements will be acceptable by all JAA states.

- *JAR-FTL*
  these are the requirements for Flight Time Limitations (FTL) which, when agreed, will be implemented by all JAA states.
- *JAR-OPS 1 and 3*
  these deal with the rules governing commercial transportation by aeroplanes and helicopters, respectively. The rules agree fairly closely with those currently in use in the UK. There are one or two minor discrepancies which will gradually be resolved.

European Union (EU)

The EU, representing 15 nations, will gradually be increasing its authority on aviation legislation. It has already affected the rules for aircraft moving within the EU by simplifying customs regulations (see Chapter 15).

Eurocontrol agreement

The members of this agreement, which includes most of the EU plus some other states such as Norway and Switzerland, are working towards a European Organisation for Air Safety. On behalf of the European Civil Aviation Conference (ECAC) of 32 states it is working to produce a 'seamless' air traffic management system for Europe.

UK aviation law documents

All UK aviation law is based on the Civil Aviation Act which enables various law documents to be prepared. To a large extent these documents follow the SARPs laid down by ICAO and will gradually be incorporating the regulations in the various JAR referred to earlier. The following documents will provide the source material for this book and should be referred to if further detail or clarification is required. Apart from the information circulars, purchase is not recommended because of the great cost involved.

- *Air Navigation Order (ANO)*
  this is the principal law document. Much of its content is similar to information in JAR-OPS 1 and 3.
- *Air Navigation (General) Regulations (ANGR)*
  these are an offshoot of the ANO and deal particularly with aircraft loading, aircraft performance, Mandatory Occurrence Reporting

(MOR), Minimum Navigation Performance Specifications (MNPS) required in certain areas such as the North Atlantic and various other lesser items. Again most of this material is covered in JAR-OPS 1 and 3.

- *Rules of the Air Regulations (R of A)*
  the name is reasonably self-explanatory, but it also includes rules for aircraft on the ground and aerodrome signals. Some of this information is also dealt with in JAR-OPS 1 and 3.

- *Civil Aviation (Investigations of Air Accidents & Incidents) (I of A)*
  the name is self-explanatory. JAR-OPS 1 and 3 only refer to the need to report accidents involving serious injury as defined in ICAO Annex 13 or death of anyone or substantial damage to aeroplanes or property.

- *UK Aeronautical Information Publication (UKAIP)*
  AIPs are an ICAO requirement (Annex 15) and form part of the total Aeronautical Information Service (AIS) integrated package (see Chapter 5). It gives details of all procedures and regulations affecting air traffic in UK airspace. It also gives notice of any variations in the UK from ICAO SARPs.

- *Aeronautical Information Circulars (AIC)*
  these are not strictly legal documents, but highlight information of particular importance to all pilots. They have the same sort of standing as the Highway Code, suggesting sensible ways to operate. Interestingly some information contained in AICs as suggestions now appear in JAR-OPS 1 and 3 as positive requirements.

- *Air Operator's Certificate (AOC)-(CAP 360)*
  this contains guidance for any operator wishing to obtain an AOC. Much of the information is now dealt with in JAR-OPS 1 and 3.

- *The Avoidance of Fatigue in Aircrews (CAP 371)*
  This sets out suggested work patterns for flight crew and cabin staff designed to prevent the onset of fatigue. It is not a legal document but an operator is unlikely to get approval for operations unless its advice is followed. This will be replaced by JAR-FTL (Flight Time Limitations).

Chapter 2
Airspace Divisions

Introduction

Practically the whole world is flown into or over by civil air traffic and so there is a very real need for there to be order and control, to a greater or lesser extent, depending on the density of the traffic. ICAO give as the objectives of the Air Traffic Services (ATS) to:

- prevent collisions between aircraft in the air and on the ground
- expedite and maintain an orderly flow of air traffic
- provide information useful for the safe and efficient conduct of flights
- notify appropriate organisations regarding aircraft in need of search and rescue aid and to assist such organisations as required.

To enable these objectives to be realised it is important to divide all airspace into areas of responsibility.

Flight Information Regions (FIR)

Over all countries the airspace is known as an FIR. Quite often it extends to surrounding sea areas. In each FIR there is an Air Traffic Control Centre (ATCC or ACC) with responsibilities for:

- control of aircraft within certain areas
- providing an efficient meteorological service
- maintaining an FIS (Flight Information Service)
- providing a communications centre
- providing an alerting service for aircraft needing assistance.

FIR boundaries are shown on all maps as in Figure 2.1.

The ICAO location indicators (ENSV and EGPX) will be needed when filing flight plans. The G in brackets indicates the type of airspace; G is uncontrolled airspace.

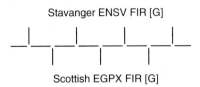

Figure 2.1 FIR boundary marking on charts.

Upper Flight Information Region (UIR)

Above FIRs there will often be UIRs. In the UK the dividing line is Flight Level (FL) 245, but quite a number of European countries use FL 195. The upper limit in the UK is FL 660 but in others (Germany and Italy, for example) it is only FL 460. There are no universal rules, but on the front of en-route charts there will usually be a table showing the limits. The area in the FIR is Lower Airspace and in the UIR Upper Airspace. UIR boundaries are shown in the same way as for FIRs, but the titling will indicate UIR.

Controlled Airspace (CAS)

Within the FIRs there will be areas where the amount of traffic makes it essential for the Air Traffic Control (ATC) to exercise authority and responsibility to ensure the safety of the traffic. These areas are designated Controlled Airspace (CAS) and under ICAO they are classified into five categories according to the degree of control exercised.

- *Class A*
  this is the highest category and in it all flights must generally be conducted under Instrument Flight Rules (IFR). This entails the pilot being Instrument Rated (IR), a flight plan must be submitted and a flight clearance obtained. The aircraft must have a certain minimum level of equipment so that it can be safely navigated in the CAS. A full ATC service is provided to all aircraft under all conditions day and night. The ATC will provide procedural separation by carefully allocating suitable Flight Levels (FL), routes and times of departure. In flight, in many cases, and certainly within the UK, radar surveillance will be carried out to ensure aircraft are complying with their clearances. In the UK practically all the airways and all the major CTAs are class A.
- *Class B*
  in the rules to be followed and the services offered this is the same as

class A, but flights under Visual Flight Rules (VFR) are permitted. In the UK, the whole of the Upper Airspace (UIR) is in this class.

- *Class C*
  the only way this differs from class B is that VFR traffic is offered separation from IFR traffic but only information on other VFR traffic on request. No class C airspace is in existence in the UK at the moment.
- *Class D*
  in this, IFR traffic is separated from other IFR traffic and traffic information is provided on conflicting VFR traffic. For VFR traffic, information is provided on all other flights so that pilots can effect their own traffic avoidance and integration. In the UK the less important Control Zones (CTR) and Control Areas (CTA) are class D.
- *Class E*
  in this, IFR traffic is separated from other IFR traffic. Separation from VFR traffic is given as far as is practical. The reason for this is that VFR traffic does not have to obtain a clearance and so ATC may not be aware of all VFR traffic. In addition, VFR traffic does not have to have radio communication facilities and so ATC may not be able to contact VFR traffic seen on its radar. In the UK there are very few class E areas.

Types of controlled airspace

There are a number of types of CAS and each will be given an ICAO classification. These classifications will be indicated on most en-route charts. In the UK the UKAIP (United Kingdom Aeronautical Information Publication) will also give these details. Figure 2.2 shows how all UK airspace, both controlled and uncontrolled, is divided.

Control Areas (CTA)

These are CAS which extend upwards from an altitude above ground level. Their lower limit may be defined as an altitude or a FL. Usually they extend upwards to the limit of lower airspace unless otherwise notified. Special cases of CTAs:

- *Airways (AWY)*
  these are comparatively narrow strips along the main air routes. They are always well marked with radio facilities and in the UK they are all class A airspace and so all flights have to be under IFR. In some countries they may not be class A and VFR flight may be permitted subject to clearance being obtained. Usually airways are never lower than 3000 ft above the ground and only extend to the limit of the lower air-

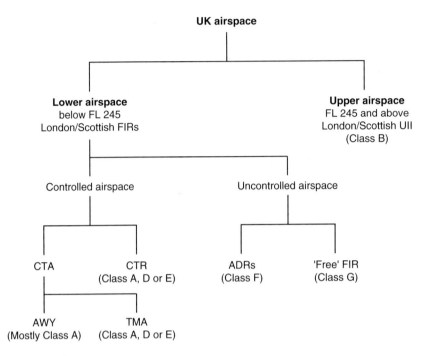

Figure 2.2 UK airspace divisions.

space (FL 245 in the UK). Above the airways there will often be an upper ATS route (classified B in the UK). Airways are designated by a letter from the phonetic alphabet plus a number – Alpha 1, Bravo 5, etc. All airways shown on charts will have Minimum En-route Altitude (MEA) shown on them and sometimes if the airway does not extend to the top of the lower airspace an MAA (Maximum Authorised Altitude) may be shown; see Figure 2.3. The letters MEA do not appear on the chart. Upper ATS routes will be shown starting with a letter U, UA1 for example. UK Airways are generally only 10 nm wide but even in some parts of Europe wider airways will be found and in areas of the world, with much less traffic, even wider ones will be met.

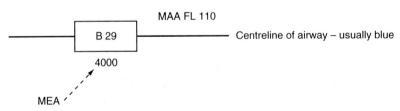

Figure 2.3 Airway showing MEA and MAA.

- *Terminal Manoeuvring Areas (TMA)*
  these are CTAs established near busy airports, where many airways converge and in which positive control will be exercised to separate all traffic. The most important TMAs will be class A. Less important ones will be C, D or E. The classification will always be noted on the en-route charts.

Control zones (CTR)

These are areas of CAS which extend from the surface to an altitude or a flight level. They will be found around important aerodromes. Entry to them always requires clearance from the controlling authority (ATC) and they will very often be class A.

Types of uncontrolled airspace

- *Advisory Service Areas (ASA) and Advisory Routes (ADR)*
  these are classified by ICAO as F. In the UK, there are only ADRs and not ASAs. In class F airspace separation can only be provided between IFR traffic that has decided to use the advisory service. The ATC will not have details of non-participating traffic either VFR or IFR and so cannot guarantee separation from this traffic. It is possible that the flight will be under radar surveillance and so warnings may be given to participating traffic of responses seen on the radar, but, with no knowledge of the intentions of this traffic, nothing more can be done. An ADR will have all the appearances of an airway on the charts, but the identification will always be followed by a 'D'. For example, W 14D.
- *The 'Free' FIR*
  all airspace that is not classes A–F, is referred to as class G. No ATC is operated here but a Flight Information Service (FIS) may be provided. In addition, some aerodromes may be able to offer a Radar Advisory Service (RAS) or a Radar Information Service (RIS). This is dealt with in Chapter 10.

Aerodromes and Aerodrome Traffic Zones (ATZ)

The international legislation relating to aerodromes is set out in ICAO Annex 14 and in 'Procedures for Air Navigation Services' (PANS RAC ICAO Doc 4444). An aerodrome is defined as any area of land or water designed, equipped, set apart or commonly used for the movement of aircraft of any type. Within the UK the following aerodromes will be encountered:

• government aerodromes
• aerodromes having an ATCU or an aerodrome flight information unit
• a licensed aerodrome having two-way radio.

and will have standard ATZs during the notified hours of watch at the aerodromes. Their dimensions are given in Table 2.1.

ATZs are not allocated a specific class of airspace but adopt the class of the airspace within which they lie.

Rather like much controlled airspace, aircraft cannot enter ATZ without permission from the ATC or, if there is no ATC, without obtaining all necessary information from the Aerodrome Flight Information Service Officer (AFISO) or the air/ground radio operator.

Table 2.1 Aerodrome traffic zone dimensions

Aerodrome type	Up to level	Radius from centre of aerodrome
Offshore installation	2000 ft amsl	1.5 nm
Longest runway 1850 m or less	2000 ft aal	2.0 nm
Longest runway over 1850 m	2000 ft aal	2.5 nm

amsl, above mean sea level; aal, above aerodrome level.

Military Air Traffic Zones (MATZ)

In the UK certain military aerodromes have MATZ to provide increased protection to their traffic. Civil flight through MATZ does not require clearance but is advisable. Within the MATZ the standard ATZ exists through which clearance is required. The dimensions of a MATZ are shown in Figure 2.4.

Types of Air Traffic Service Units (ATSU)

• *Air Traffic Control Centre (ATCC)*
  this combines the functions of an Area Control Centre (ACC) and a Flight Information Centre (FIC). Within its area of responsibility (FIR and/or UIR), it provides the necessary services for all aircraft whether inside or outside CAS. An ACC is the more usual term in other countries and its purpose is to provide ATC to IFR flights. An FIC provides

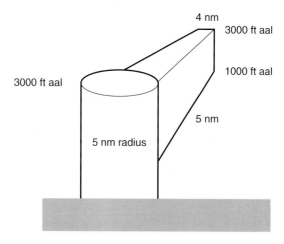

Figure 2.4 Dimensions of UK Military Air Traffic Zone (aal, above aerodrome level).

Flight Information Service (FIS) and alerting service to all aircraft. Figure 2.5 shows the services that an ATCC provides.
- *Zone Control Unit (ZCU)*
  larger control zones are usually dealt with by the ATCC, but smaller ones may have their own ZCU dealing with aircraft within their zone.

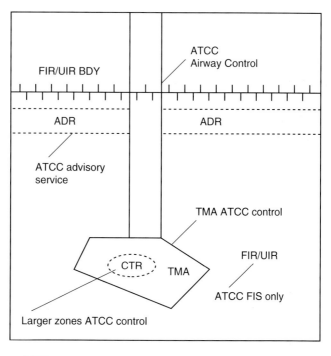

Figure 2.5 ATCC responsibilities..

A zone controller does not control the take-off and landing phases of flights. This is dealt with by the aerodrome controllers.

• *Aerodrome control*

this may be divided into various responsibilities depending on the amount of traffic being handled:

 – Ground Movement Controller (GMC): at larger airfields the GMC will be responsible for all ground movements

 – Aerodrome Control Unit (ACU or tower): provides control service at the aerodrome. At smaller aerodromes it may handle the ground movements and the approach responsibilities

 – Approach Control Unit (ACU): this provides control service to aircraft taking-off and landing. At smaller airports it may also provide the Zone Control in smaller zones.

Some aerodromes may have even more subdivisions of responsibility. There may, for example, be a departure, as distinct from an approach controller.

Changes of control throughout a flight

An aircraft taking off from an aerodrome within a zone will, in sequence, deal with:

• Ground Movement Control if there is one
• Aerodrome Control
• Approach Control
• ATCC during flight in the TMA, if this exists
• ATCC airways controllers
• ATCC for TMA at destination
• ATCC for Zone at destination
• approach controller at destination
• aerodrome control (tower)
• Ground Movement Control, if there is one.

Chapter 3
Visual Flight and Instrument Flight Rules (VFR and IFR)

Introduction

In the previous chapter on controlled and uncontrolled airspace reference has been made to VFR and IFR. Here we look at the rules governing these types of flight.

Flight conditions required for VFR operations

Table 3.1 is from the appendix to JAR-OPS 1.465. It gives the minimum conditions of visibility and distance from cloud for the pilot to maintain VFR. The table in the UKAIP En-route section (ENR) is essentially the same as Table 3.1.

Table 3.1 Minimum conditions for flying VFR

Airspace class	B	C, D and E	F and G	
Levels	–	–	Over 900 m (3000 ft) or above 300 m (1000 ft) over terrain, whichever is greater	900 m (3000 ft) and under or above 300 m (1000 ft) over terrain, whichever is greater
Distance from cloud	Clear of cloud	1500 m horizontally, 300 m (1000 ft) vertically	Clear of cloud and in sight of surface	
Flight visibility	8 km at and over 3050 m (10 000 ft) amsl* 5 km below 3050 m amsl‡		5 km†	

*When the height of the Transition Altitude (TA) is less than 3050 m amsl, FL 100 should be used instead of 3050 m.
†Cat A and B aeroplanes (smaller, more manoeuvreable aircraft) may be operated when flight visibility is down to 3 km provided ATS permits the use of a flight visibility less than 5 km and the circumstances are such that the risk of meeting other aircraft is low and the Indicated Air Speed (IAS) is under 140 kt.
‡If helicopter and flying at 1000 m or below – clear of cloud and sight of the surface.

VFR flights taking-off or landing in a Control Zone (CTR)

According to the UKAIP ENR, unless given clearance by ATC, VFR flights shall not take-off or land in a Control Zone when the ceiling is less than 1500 ft (450 m) above aerodrome level (aal) or when the ground visibility is under 5 km. Cloud ceiling is the height aal of the lowest part of the cloud visible from the aerodrome obscuring over half the sky.

The UK does not permit VFR flights in certain CTA and CTR – these are notified in the UKAIP.

Restrictions on VFR flights in UK

The following restrictions are laid down if it is required to fly VFR in the UK:

- *Night flights*
  generally night flights are to be conducted under IFR but, if VFR is permitted, it will be in accord with ATS instructions.
- *Flights above FL 290*
  not permitted where a vertical separation minimum of 1000 ft is prescribed.
- *Transonic and supersonic flights*
  only if authorised by ATS.
- *Above FL 200 in Notified Control Areas (CTA) or Control Zones (CTR)*
  not permitted.
- *Over congested areas or open-air assemblies of persons*
  height must be 1000 ft above the highest obstacle within 2000 ft of the aircraft unless when taking-off or landing or with permission of the appropriate authority.

Pilot's responsibilities in VFR flight

Under VFR the pilot is responsible for avoiding any obstacles including other aircraft. With permission, the pilot may fly in any Controlled Airspace (CAS) apart from class A airspace (airways and main Control Areas), but in these cases the clearance given must be followed as regards route and altitudes. In the UK, the pilot does not have to adopt any particular rule for cruising level, although this is not necessarily the case under ICAO rules which apply in many countries. In the UK, flights above 3000 ft above mean sea level (amsl) are recommended to adopt the rules for IFR flights.

Changing from VFR to IFR

If there is no flight plan for the flight and the aircraft is not in any form of CAS there is no problem but in other cases the correct procedure must be followed for IFR flight, which will probably require the filing of a flight plan and the obtaining of a clearance.

Special VFR in the UK (SVFR)

- *Purpose*
  to enable flight in a CTR when IFR flight would normally be required because of the weather conditions, it being night or it is class A airspace, which always requires IFR flight.
- *When given*
  normally only to small (under MSTOM of 5700 kg), when conditions permit without hindrance to IFR traffic. Visibility must be over 1800 m and the cloud ceiling over 600 ft unless the pilot has an Instrument Rating (IR) or an Instrument Meteorogical Conditions (IMC) rating and the visibility is at least 10 km.
- *Instructions given*
  will include a flight route and possibly a height limit. The conditions should enable the pilot to check the flight path and keep clear of obstacles. The pilot must keep clear of cloud and in sight of the surface. Also, the pilot must not infringe any ATZs without prior permission.
- *Low flying rules*
  these must be complied with apart from the rule requiring that in congested areas the flight must be at least 1500 ft above any obstacle within 600 m (2000 ft). The rule requiring after engine failure to be able to land clear of the congested area must be complied with. Single-engine aircraft, therefore, can never get SVFR clearances over large metropolitan areas. There are, however, special rules for helicopters over London.
- *Service given by ATS*
  separation will be provided from all other aircraft.
- *Flight plans*
  full plans are not needed, but brief details of the flight required. If the destination aerodrome is to be notified a full flight plan must be filed. This can be done by radio.

IFR flight in UK

- *Minimum height for IFR*
  unless on a notified route, taking-off or landing or authorised by ATS or flying below 3000 ft amsl and clear of cloud and in sight of the surface,

the pilot must ensure a clearance of at least 1000 ft above the highest obstacle within 5 nm.

- *Outside controlled airspace*

special rules are laid down for the pilot to observe. The ICAO rules are used in most countries but the UK and some places overseas (mostly those that were once under British rule) have a different set of rules for the lower levels.

- *Quadrantal rules*

in Europe the following rules are peculiar to the UK.

IFR flight levels outside CAS and above 3000 ft amsl and below 24 500 ft amsl

Magnetic track	Cruising levels
000° to 089°	Odd thousands of feet (FL 30, 50, 70, ... , 230)
090° to 179°	Odd thousands + 500 ft (FL 35, 55, 75, ... , 235)
180° to 269°	Even thousands of feet (FL 40, 60, 80, ... , 240)
270° to 359°	Even thousands + 500 ft (FL 45, 65, 85, ... , 225)

- *Semi-circular rules*

IFR flight levels outside CAS and above 24 500 ft amsl

Magnetic track	Cruising levels
000° to 179°	FL 250, 270, 290
	FL 300 is not used
	FL 330, 370 and every 4000 ft
180° to 359°	FL 260, 280
	FL 300 is not used
	FL 310, 350 and every 4000 ft

- *ICAO Semi-circular rules*

IFR flights outside CAS from 1000 ft amsl

Magnetic track	Cruising levels
000° to 179°	FL 10, 30, 50, ... , 290
(ODDs)	FL 300 is not used
	FL 330, 370 and every 4000 ft
180° to 359°	FL 20, 40, 60, ... , 280
(EVENs)	FL 300 is not used
	FL 310, 350 and every 4000 ft

In addition, there are allocated levels for VFR flight which are 500 ft above the IFR levels up to 28 500 ft and then over FL 300, 1000 ft above the IFR levels.

- *IFR levels inside CAS*
  the levels to be flown will be allocated by ATC. In general they will adhere to the ICAO semi-circular rules and will be referred to as ODDs (easterly routes) or EVENs (westerly routes). En-route charts will indicate where the semi-circular rules are not likely to be observed by annotating the airway centre line O> or E> for odds or evens. It can be accepted that, if the airway is usable in the reciprocal direction, the reverse rule will apply in the opposite direction.
- *IFR inside UK CAS*
  the rules to be followed are in ICAO Annex 2 except for minor differences:
  - all flights will require a flight plan to be submitted and a clearance obtained
  - the aircraft must be equipped with suitable instruments and navigation aids appropriate to the route being flown
  - the cruising levels flown will be as authorised in the clearance, as will the route
  - the pilot will require an Instrument Rating (IR).

Reduction of Vertical Separation Minima (RVSM)

A yellow Aeronautical Information Circular (AIC) describes the move towards reducing the vertical separation minima in certain selected areas such as the North Atlantic, where the separation between FL 310 and 390 has been reduced to 1000 ft, instead of 2000 ft. From November 2001 it is planned to introduce RVSM in all the European Conference for Civil Aviation (ECAC) airspace.

JAR-OPS 1 1.243 refers to RVSM and forbids operating in areas where RVSM have been agreed unless approved to do so by the Authority. Such approval will only be granted (JAR-OPS 1.872) if the aircraft has:

- two independent altitude measuring systems
- an altitude alerting system
- automatic altitude control system
- a secondary surveillance radar (SSR) transponder with an altitude reporting system that can be connected to the altitude measuring system in use for altitude keeping.

Chapter 4
Altimeter Setting Procedures

Introduction

The altimeter measures air pressure and relates this to the pressure set on a subscale, indicating the equivalent distance in feet represented by the difference between the pressures. There are three settings in general use and these are described below. They are known by the Q groups, which date back to the use of wireless telegraphy which had a whole range of three-letter Q groups, many of which survive to this day but in their spoken form.

QFE

This is the pressure at aerodrome reference level. The aerodrome reference level is the highest point within the landing area. With QFE set, the altimeter will normally read 0 ft on landing. If the precision landing runway is 7 ft or more below the airfield elevation the QFE for the landing threshold will be passed so that, when this value is set, the altimeter will read 0 ft on touch-down.

QNH

This is a derived sea-level pressure calculated so that, when set on the subscale, the altitude indicated will be that of the aircraft above mean sea-level. On landing, with QNH set, the altimeter should read the aerodrome elevation. As it gives a reasonable indication of altitude above mean sea-level it is a useful setting for checking terrain clearance. As a better guidance, ATC will give a Regional QNH which is the lowest QNH likely to prevail in an Altimeter Setting Region for each hour. Pilots setting Regional QNH will get an altitude which will give them a reasonable indication of whether they are clear of the surrounding high ground in that region.

Standard setting

This is the setting of the sea-level pressure of the International Standard Atmosphere (ISA) – 1013.2 mb.

When set, the altitude showing on the altimeter is the altitude the aircraft would have to be at in the standard atmosphere to obtain the pressure the aircraft is at present experiencing. This is referred to as the Pressure Altitude (PA) of the aircraft. When expressed in hundreds of feet it gives the Flight Level (FL). For example, when required to fly at a FL of 120, the pilot, with 1013 mb on the subscale, will fly to keep the altimeter reading 12 000 ft. This technique is used to ensure that all traffic in the same airspace will be separated by approximately the amounts indicated by their altimeters, which will all be set to 1013 mb.

Altitudes and heights

- *Height*
  vertical distance of a point from a datum. The height of an aerial mast would be how far it projects above the ground.
- *Altitude*
  vertical distance of an aircraft above mean sea-level (amsl).
- *Elevation*
  the same as altitude but used for airfields.
- *Transition Altitude (TA)*
  is the altitude in the vicinity of an aerodrome below which the vertical position of an aircraft is controlled by reference to altitude. In the UK, where there are no aerodromes at a great elevation, the usual TA is 3000 ft. Within some CTR and within or below some CTA, which are notified in the UKAIP, higher levels (4000–6000 ft) will apply. The TA is always noted on Instrument Approach Charts (IAC).
- *Transition Level (TL)*
  this is the lowest FL available which is physically (not necessarily numerically) above the TA.

Figure 4.1 illustrates the different altitudes and flight levels.

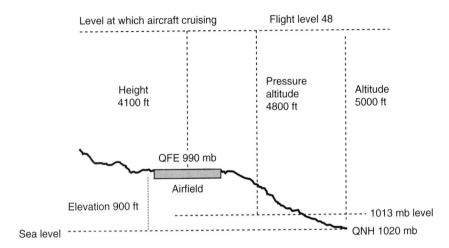

Figure 4.1 Altitudes and flight levels.

Levels to fly in VFR

- *Inside CAS*
  the levels will be allocated by ATC. Generally they will adhere to the ICAO semi-circular rules and will be referred to as ODDs (easterly routes) or EVENs (westerly routes). En-route charts will indicate when these rules are not likely to be observed by annotating the airway centre line with O> for ODDs or E> for EVENs.
- *Outside CAS generally*
  the ICAO semi-circular rules will apply
- *Outside UK CAS*
  the quadrantal rules will apply. Remember that Advisory Routes (ADRs) are not in CAS.

Altimeter setting procedures

- *For take-off within CAS*
  at least one altimeter to be set to aerodrome QNH. Reference to vertical position is in terms of altitude up to transition altitude (TA). However, if cleared to a FL before reaching the TA, FL may be used (altimeter reset to 1013 mb) unless intermediate altitude reports are required.
- *En-route above the transition level*
  the standard setting (1013 mb) and FLs will be used. For terrain clearance purposes use should be made of the Regional QNHs.

- *Nearing destination*

  when an aircraft is cleared to descend from a FL to an altitude prior to commencing an approach to landing the aerodrome QNH will be passed. The pilot will change to QNH on vacating the FL, unless further FL vacating reports are required by ATC.

- *Final approach*

  when established on the final approach, ATC will assume QFE is being used if carrying out a radar approach. Some operators use the QNH for final approach and the ATC will ensure that QNH is named in their messages. If landing at aerodromes below a TMA or CTA and below TA, pilots should use the aerodrome QNH until in the circuit when QFE may be used.

Figure 4.2 illustrates these altimeter setting procedures and Figure 4.3 the procedure when landing at an aerodrome under a TMA.

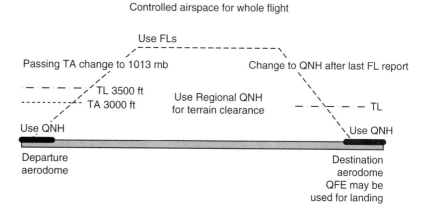

Figure 4.2 Altimeter settings for flight wholly in CAS.

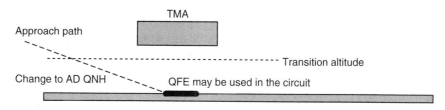

Figure 4.3 Landing at an aerodrome beneath a TMA.

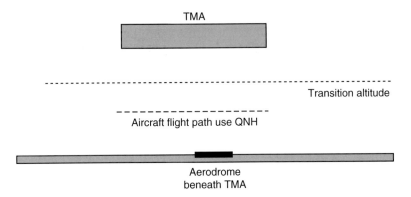

Figure 4.4 Flight altimeter setting to use when flying below a TMA.

- *For take-off outside CAS*
  take-off and initial climb may be done using any setting.
- *En-route outside CAS*
  at and below 3000 ft any settings may be used but, if reporting vertical position, altitude (QNH setting) must be used. Flying above 3000 ft the same rules apply as for flying in CAS. If flying below a TMA or CTA, use the QNH of an aerodrome beneath the area; see Figure 4.4.
- *Approach and landing outside CAS*
  use the same procedures as within CAS.
- *Missed approaches*
  altimeter setting for the final approach may still be used but vertical position must be reported as an altitude (QNH setting).
- *Pre-flight altimeter checks*
  in the UK, so that the pilot can check his altimeters while on the apron (which is the designated place for doing this), the apron elevations will be shown in the flight clearance office.

Chapter 5
Aeronautical Information Service (AIS)

Introduction

Annex 15 of the ICAO Convention deals with Aeronautical Information Services to be provided by member states. A central feature of this service is the Aeronautical Information Publication which forms part of the Integrated Aeronautical Information Package. The layout of this package is summarised in Table 5.1.

Keeping the UKAIP up to date

In accordance with Annex 15, the UKAIP is part of an integrated package that is supported by the following amendments and supplements:

- *Regulated System for Air Information (AIRAC) amendments*
  these are published 4-weekly and give 6 weeks advance notice of changes that will all occur on one particular date. Pilots will know that only in exceptional cases will changes occur on other dates. These amendments take the form of replacement pages for the UKAIP.
- *AIP Amendments (Non-AIRAC)*
  these are also published 4-weekly and deal with non-operationally significant changes.
- *AIP Supplements (formerly NOTAMs II)*
  these are issued fortnightly and cover items of a temporary nature.

NOTAMs (Notices to Airmen)

All other information not covered by AIP supplements or amendments will be issued as a NOTAM on the Aeronautical Fixed Telecommunication Network (AFTN). These used to be referred to as NOTAMs I and will cover operationally significant changes that need to be introduced at

Table 5.1 The Integrated Aeronautical Information Package

NOTAM and Pre-flight Information Bulletins (PIB)	AIP including Amendments and Supplements	Aeronautical Information Circulars (AIC)
Part 1 GENERAL (GEN) **GEN 0** Preface	**Part 2 EN-ROUTE (ENR)** **ENR 0** Preface	**Part 3 AERODROMES (AD)** **AD 0** Preface
GEN 1 National Regulations and requirements including entry, transit and departure procedures. Notifications where UK departs from SARPs (Chapter 1)	**ENR 1** General rules and procedures. Types of airspace, Royal flights, restrictions and hazards, holding, approach and departure procedures, altimeter settings, air traffic flow. Management (AFTM), flight plans, interceptions, unlawful interference, AIRPROX.	**AD 1** Aerodromes/heliports – introduction. General conditions including AOM, use of government aerodromes, aerodrome closures, Cat II/III operations, snow plan, rescue and fire organisation
GEN 2 Tables and codes abbreviations, chart symbols, ICAO location indicators, radio navigation aid codes, sunrise/set tables	**ENR 2** ATS airspace. Details of FIR, UIR, TMA and other regulated airspace	**AD 2** Aerodromes. Alphabetical list giving full details including all services available, aerodrome plans, routes including SIDs and STARs, radar vectoring, Instrument Approach Charts (IAC)
GEN 3 Services. AIS, charts, ATS, communications, meteorology, search and rescue	**ENR 3** ATS routes. Details of lower and upper ATS routes. Area navigation routes. Helicopter routes. En-route holding	**AD 3** Heliports. Same as for AD 2
GEN 3 Services. AIS, charts, ATS, communications, meteorology, search and rescue	**ENR 3** ATS routes. Details of lower and upper ATS routes. Area navigation holding	**AD 3** Heliports. Same as for AD 2
GEN 4 Charges for aerodromes, heliports and air navigation services	**ENR 4** Radio navigation aids and systems. En-route aids. Special navigation systems. Name-code designators for significant points. Aeronautical ground lights – en-route	
	ENR 5 Navigational warnings	
	ENR 6 En-route charts	

short notice. They will usually be followed up by AIP amendments or supplements.

- *Trigger NOTAMs*
  operationally significant changes already announced in amendment or supplements will additionally be 'flagged up' by Trigger NOTAMs giving the effective date and AIP reference. This ensures that the information will appear in Pre-flight Information Bulletins (PIB).
- *System NOTAMs*
  as required by ICAO, the UK now uses the 'System' NOTAM format. This is designed so that information can be extracted automatically by computers and incorporated in PIBs.

Aeronautical Information Circulars (AIC)

These are issued monthly and contain administrative, operational, safety matters and amendments to UK airspace restriction charts. They are to draw the airman's attention to new matters that are arising. They are colour coded as follows:

- *white*   administrative
- *yellow*   operational including ATS facilities and requirements
- *pink*   safety
- *mauve*   amendments to airspace restriction charts
- *green*   maps and charts information.

AICs lapse after 5 years but, if still considered important, may be re-issued. The AICs are part of the AIS integrated package.

UK Aeronautical Information Service (AIS)

This is responsible for the collection and dissemination, to all interested parties, of information necessary for the safety and efficiency of air navigation throughout the UK including all airspace for which the UK is responsible. AIS is located at London (Heathrow) Airport and has the following sections:

- NOTAM office
- publications (AIP, etc.)
- Pre-flight Information Bulletins
- foreign records library
- foreign library enquiry desk.

SNOWTAMS

These are a special type of NOTAM which give information on the condition of movement areas and the current state of snow clearance. They are issued at least every 24 hours or when significant changes take place. In the UK a white AIC is issued annually which gives:

- aerodromes where standard clearance will be carried out
- equipment held at each aerodrome and the type of clearance
- height and distance of snow banks permitted at each aerodrome
- method of assessing braking action
- authority to contact for current information
- any local deviation from standard practice.

Pre-flight Information Bulletins

These are prepared from NOTAMs, usually by computer, and give details of all current NOTAMs for a particular area. They are available daily at main aerodromes and also by post, facsimile or screen-based services.

Chapter 6
Aerodromes – General

Introduction

ICAO Annex 14 deals with:

- *Volume 1* – aerodrome design and operations
- *Volume 2* – heliports.

This chapter deals with:

- limitations of use
- operations from contaminated runways
- definitions
- ground markings
- airfield approach lighting
- Runway Visual Range (RVR)
- aeronautical ground light beacons.

Limitations on the use of aerodromes

- *General*
  civil aircraft are not permitted to land at any aerodrome not listed in the UK AIP except in an emergency. Aircraft may land at unlisted aerodromes with permission of the appropriate authority. The permission is only granted to land at these and disused airfields in exceptional circumstances such as medical emergencies and tasks of national importance.
- *Government aerodromes*
  permission is required, before take-off, to use a government aerodrome during published working hours. It is government policy to encourage the use of these for internal flights.
- *Ordinary aerodrome licence*
  in this case the use is only with the prior permission of the owner or manager.

- *Unlicensed aerodrome*
  use only with prior permission of owner.
- *Military aerodrome*
  the use is similar to that for government aerodromes, but checks should be made. Foreign aircraft may only use these if they are so diverted by ATC.

Snow clearance

SNOWTAMs have already been referred in Chapter 5. The following points should be noted:

- *Aerodrome responsibility*
  clearing the snow with the operational runway being top priority and measuring and reporting the aerodrome state.
- *Clearance technique*
  if possible all runways should be completely cleared. Slush and standing water to be cleared if more than 3 mm deep. Non-toxic chemicals only should be used. Grit may be used to improve braking action.
- *Measurement and reporting*
  special equipment will be used to assess the braking conditions on ice, snow or slush as GOOD, MEDIUM or POOR and this will be passed to the pilots. Wet runways are similarly reported and indication of whether aquaplaning is likely to occur. The snow density on the ground together with the type of snow is also reported.

Operating from contaminated runways

In general this should be avoided by diverting, if airborne, or delaying departure before take-off. Water on a runway is reported to pilots as DAMP, WET, WATER PATCHES or FLOODED. A pink Information Circular (AIC) deals at length with this subject. It points out the following effects of over 3 mm of water or 10 mm of dry snow:

- *Additional drag*
  retardation effects on the wheels and spray impingement drag.
- *Power loss*
  due to spray ingestion or impingement.
- *Reduced braking*
  aquaplaning.
- *Directional control problems*
- *Possible structural damage*

The general advice for operating from contaminated runways is as follows:

- do not take-off if water, slush or wet snow is over 15 mm, or dry snow is over 60 mm or very dry snow is over 80 mm
- avoid using reverse thrust, when manoeuvring prior to take-off, to avoid contamination of wing leading edges
- consider all aspects when selecting flap/slat configuration and ensure that all field length performance corrections are made
- keep fuel carried to a minimum
- all braking devices should be serviceable and the tyres in good condition
- use maximum take-off power
- avoid tail winds and crosswinds over 10 kt
- use normal rotation and take-off safety speeds
- check all de-icing necessary has been done
- use maximum runway distance available.

Aerodrome definitions

- *Elevation*
  the altitude above mean sea level (amsl) of the highest point in the landing area. It is given in the aerodrome (AD) section of the UKAIP and on all Instrument Approach and Aerodrome Charts.
- *Alternate*
  an aerodrome, selected prior to take-off, to which the flight may proceed when landing at the destination aerodrome becomes inadvisable.
- *Supplementary aerodrome*
  an aerodrome to be used in special cases when unable to use the planned destination or alternate.

Runway distances (see Figure 6.1)

These are defined in JAR-OPS 1.480:

- *Declared distances*
  these are the official published values for the various distances described below.
- *Take-off Run Available (TORA)*
  take-off run available – it will often be the runway length.

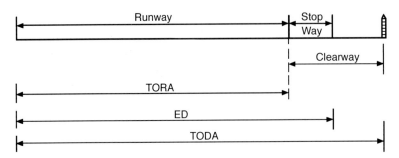

Figure 6.1 Runway lengths for take-off.

- *Accelerate–Stop Distance Available (ASDA)*
  the TORA plus the stopway available.
- *Take-Off Distance Available (TODA)*
  the length of the TORA plus the length of the clearway available. It must never exceed 1.5 times the TORA.
- *Landing Distance Available (LDA)*
  is the length of the runway (or surface if unpaved) suitable for the ground landing run of the aircraft commencing at the visual threshold markings or lights. If the landing area is unpaved, the LDA is measured from where a 1:20 (approximately 3°) approach path which is free from obstructions cuts the surface.
- *Stopway*
  is the area beyond the TORA suitable for the ground run of an aircraft which is decelerating.
- *Clearway*
  is the area from the end of the TORA selected and prepared as suitable for the aircraft to make its initial climb to a specified screen height (often 35 ft with modern aircraft) within which there are no upstanding obstacles of any height.

Performance categories – JAR-OPS 1.470

Under JAR-OPS 1 Public Transport Aircraft will be classified as follows:

- *Class A*
  multi-turbo propeller aircraft with approved passenger seating of over nine or a maximum take-off mass (MTOM) of over 5700 kg.
- *Class B*
  propeller-driven aircraft with passenger seating capacity of nine or fewer and a MTOM of 5700 kg.

- *Class C*
  aircraft with reciprocating engines and with passenger seating capacity of greater than nine or a MTOM exceeding 5700 kg.
- *Small multi-engined aircraft*
  seating more than nine passengers but with MTOM under 5700 kg may operate under alternative conditions to class A, but not less restrictive than class B.
- *Other aircraft such as supersonic and seaplanes that do not fit into the above classes*
  will be dealt with by the authority so that they have approved performance standards that ensure a safe operation.

Operator's responsibilities for adequate performance – JAR-OPS 1.475

- An operator of a public transport aircraft shall ensure that before take-off the mass is not over that which will ensure a safe operation appropriate to the classification of the aircraft throughout the flight and, in class A, even in the event of engine failure at any point on the flight.
- The approved performance data given in the Aeroplane Flight Manual (AFM) are being used for all the calculations.
- Account to be taken of the aeroplane configuration, environmental conditions and the operation of systems such as air conditioning or anti-icing which have an adverse effect on the performance. Damp runways unless they are grass may be considered to be dry.

General principles of performance A calculations

The majority of public transport operations will be in class A and, as such they are expected to be safe if an engine fails at any time during the flight. The operator must ensure that:

- *During take-off*
  the aircraft performance allowing for engine failure at any time is such that the Take-off Run Required (TORR), the Accelerate–Stop Distance Required (ASDR) and the Take-off Distance Required (TODR) to reach a specified height will all be less than the TORA, ASDA and TODA, respectively.
- *En-route*
  the aircraft at the expected mass at the time of an engine failure should be able to reach a safe landing at a suitable aerodrome.

- *Landing*
  according to calculation the landing distance required (LDR) should not exceed the LDA at the destination and alternate aerodromes.

Civil Aviation Publication (CAP) 637

This gives details of all visual ground aids, lighting, paved runway markings used in the UK.

Taxiway markings (ICAO Annex 14)

The holding position at a runway entry point is a four-line marking (Figure 6.2a). If it is necessary to provide separate holding points for visual and instrument conditions an additional ladder-shaped marking will be provided further away from the runway (Figure 6.2b).

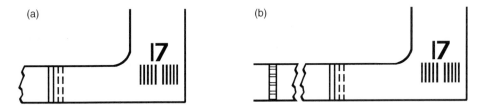

Figure 6.2 (a) Holding position; (b) holding for instrument conditions.

Runway markings at CAA aerodromes

- All runways have centreline, threshold and runway identification marks (Figure 6.3).
- Thresholds may be displaced for various reasons, such as obstructions on the final approach path (Figure 6.4).

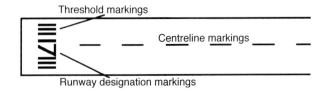

Figure 6.3 Basic runway markings.

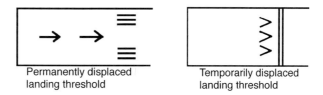

Figure 6.4 Displaced thresholds.

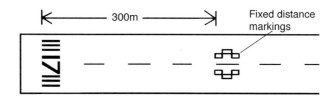

Figure 6.5 Fixed distance markings.

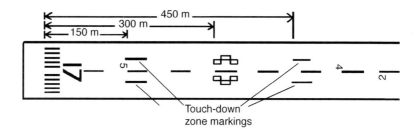

Figure 6.6 Markings on precision approach runways.

- All non-instrument runways over 1100 m long and without Visual Approach Slope Indicators (VASI) and all instrument runways have an additional 'Fixed Distance Marking' at 300 m from the threshold (Figure 6.5).
- On precision approach runways, Touch-Down Zone (TDZ) markings 150 m apart are used up to at least 600 m (Figure 6.6).

Approach lighting

The lighting arrangements at any particular aerodrome are given in the aerodrome (AD) section of the UKAIP. Apart from being able to judge the approach path from the appearance of the runway lights, the pilots may also have visual slope guidance indicators:

- *Visual Approach Slope Indicators (VASI)*
  these consist of sets of three red and white lights, the pattern of which, when observed by the pilot on the final approach, indicates when he is on the correct glide path (Figures 6.7 and 6.8).
- *Low Intensity Approach Slope Indicators (LITAS)*
  these consist of sets of two red and white lights (Figure 6.9).
- *Precision Approach Path Indicator (PAPI)*
  this is an improvement on VASI giving sharper definition. In the aerodrome details in the AIP the Minimum Pilot Eye Height over the

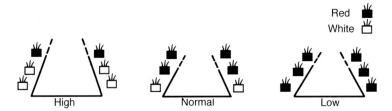

Figure 6.7 Three-bar VASI.

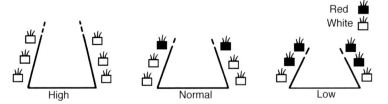

Figure 6.8 Three-bar VASI for large eye/wheel characteristics.

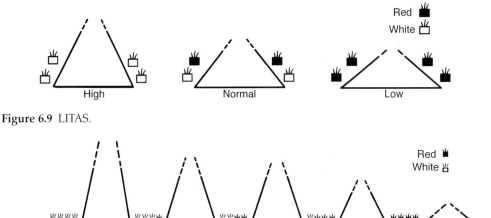

Figure 6.9 LITAS.

Figure 6.10 PAPI.

Threshold (MEHT) may be shown, being the lowest height over the threshold of the on-slope indication (Figure 6.10).

Runway Visual Range (RVR)

At some aerodromes in the AD section of the UKAIP there is a system of passing RVRs to aircraft when the visibility falls below 1500 m. RVR is one of the restrictions on aircraft landing or taking off (see Chapter 16). RVR observations are taken on the Instrument Runway in use which is normally equipped with beamed High Intensity Runway Lighting (HIRL) and also, as necessary, range markers. Range markers are triangular in shape painted black and white with the black side towards the runway. They are placed at 500 m and then at 100 m intervals from the beginning of the runway.

Measuring RVR

An observer 76 m from the centreline of the runway and as close to the landing area as possible counts the visible lights or markers that can be seen. This is then converted into the visibility that a landing pilot may expect when at a height of 5 m. At major airports Instrumented Runway Visual Range (IRVR) is available. The equipment positioned along the side of the runways provides ATS with the measured values automatically. It operates from 0 to 1500 m. At some aerodromes three IRVR may be reported and classified as A (touchdown), B (mid-point) and C (end of runway). The pilot must always consider the least of these when considering whether it is safe to land (Chapter 16).

Aeronautical Ground Lights (AGL)

The following AGL may be installed at UK aerodromes and details will be found in the en-route section (ENR) of the UKAIP:

- *Identification beacons*
  these flash a two-letter identification code in Morse. Green at civil aerodromes and red at service aerodromes.
- *Aerodrome beacons*
  usually only found at civil aerodromes without identification beacons. Originally these showed an alternating white and green light but the newer ones use a white strobe light.

Light aircraft performance

A pink Aeronautical Information Circular (AIC) deals with the take-off, climb and landing performance of these. It gives valuable guidance on all the factors that should be considered before take-off.

SNOCLO

When an aerodrome is unusable for take-offs and landings due to snow conditions the spoken word 'SNOCLO' is added to the weather broadcasts of aerodrome meteorological information (VOLMET). See Chapter 12.

Chapter 7
Flight Separation, Flight Planning, Carriage of Radio Equipment

Introduction

The ICAO agreed procedures under these headings are given in the Procedures for Navigation Services – Rules of the Air and Air Traffic Services (PANS-RAC) and are complementary to the Standards and Recommended Practices (SARPs) in ICAO Annex 2 – Rules of the Air and Annex 11 – Air Traffic Services. These are also supplemented by the Regional Supplementary Procedures (ICAO DOC 7030). Most of the information will also be found in the UKAIP, the UK Air Navigation Order and related documents and to a lesser extent in JAR-OPS 1. Useful information will also be found in Jeppesen Airways manuals.

Flight separation (Figure 7.1)

In order to prevent collisions Air Traffic Control (ATC) provides separations to aircraft under its control. Separation can be divided into vertical and horizontal types. The horizontal is further divided into:

- lateral (sideways)
- longitudinal (along the aircraft's axis)
- radar.

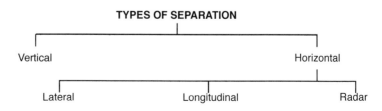

Figure 7.1 Types of separation.

Vertical separation (see Chapter 2)

This is provided by allocating different flight levels to aircraft on recipro-cal tracks. Outside Controlled Airspace (CAS) the responsibility for this will be with the pilot, but inside CAS Air Traffic Control (ATC) will allocate levels often following the semicircular rules. However, it cannot be assumed that this will always be the case. Below FL 300 the aim will be to keep a 1000 ft separation and above this 2000 ft.

Lateral separation

This can be done either by:

- ensuring aircraft leaving the same place are dispatched on diverging tracks. Account will be taken of the accuracy of the navigational tech-niques being used or
- clearing aircraft on geographically separated tracks when aircraft have got good navigational equipment. This procedure is used in the North Atlantic.

Longitudinal separation

This can be done by despatching aircraft from the same place at suitable time intervals and ensuring the aircraft are all moving at similar speeds. This latter technique is used in the North Atlantic.

Radar separation

This is done allocating different tracks determined by radar. The separa-tion minima are:

- basically 5 nm (as recommended by ICAO)
- at the discretion of the authority, when within 40 nm of the radar and below FL 245, 3 nm
- both the above assume both primary and secondary radar are available. If only secondary is available it will be 8 nm.

Air Traffic Flow Management (ATFM)

Within the ICAO (EUR) Region a centralised ATFM service is set up to optimise the air traffic system capacity available. This is provided by the Eurocontrol Central Flow Management Unit (CFMU) in Brussels. The day-to-day ATFM activity, in the airspace of the ECAC states, except Iceland, is managed by Flow Management Positions (FMP) at the Area Control Centres (ACC) and Flight Data Operations (FDO) which deal with Flight Plans. The AFTM Service has two main phases of activity:

- *Strategic*
  taking place 24 hours before the flight. Attention will be paid to possible overloading of the system and planning remedial action.
- *Pre-tactical and tactical*
  from the end of the strategic activity until the actual handling of the flight by ATC. Activities include checking the air traffic demand and amending the remedial action already planned to ensure the system is not overloaded.

The ATS and ATFM

Three FMP are set up in the UK to liase between ATC, Aircraft Operators (AO) and the CFMU. The ATC will be responsible for checking that aircraft comply with the departure slots (times) issued by the Central Executive Unit (CEU). ATC has the following responsibilities:
- *Departure slots (CTOT)*
  ensuring aircraft comply with the time departure slots issued by the CEU. Flights not doing so shall be denied start-up clearance.
- *Taxiing delays*
  short taxiing delays shall not prevent an aircraft's departure.

Responsibility of Aircraft Operators (AO)

AO shall inform themselves of, and adhere to:

- general AFTM procedures including flight plan filing
- strategic ATFM measures (e.g. Traffic Orientation Schemes)
- current ATFM measures (e.g. those applicable to the day of operation which will have been given in an ATFM Notification Message)
- departure slots (CTOT) issued by the CEU.

ATFM process

The acceptance of a flight plan within the Integrated Flight Plan System (IFPS) will provide all the information for the provision of a CTOT slot if the flight enters a regulated area. The flight plan should be filed to the IFPS at least 3 hours before the Estimated Off-Blocks Time (EOBT).

CEU ATFM process

This starts with filing of the flight plan 3 hours before the EOBT. At EOBT –2 hours the CFMU will send a Slot Allocation Message (SAM) giving the Calculated Take-Off Time (CTOT) to the aerodrome of departure and AOs via the Aeronautical Fixed Telecommunication Network (AFTN) or Société Internationale de Telecommunications Aeronautiques (SITA). Revisions to, or cancellations of, CTOT may be by the CEU, the AO or an FMP/ATC units on behalf of the AO. If it is desired to bring the EOBT earlier than the time given in the flight plan, a cancel (CNL) message is needed followed by a new flight plan 5 min later.

Permitted slot tolerance

ATC have a slot tolerance of –5 to +10 min in order to best manage departures. Operators must start in good time to taxi for the active runway to achieve the slot time and not use the tolerance for a late departure from the apron.

Inbound flow management

In congested terminal areas, short duration holding can often occur at short notice to obtain optimum airspace and runway use. In the UK, Area Control Centres (ACC) will not issue Expected Approach Times (EAT) when the terminal area delay is likely to be under 20 min. Similarly, when Approach Control is using radar sequencing they do not issue EATs for delays under 20 min.

When to file a flight plan

A flight plan may be filed for any flight and most aircraft operators will always do so. Flight plans must be filed for all flights:

- within controlled airspace that are being conducted under IFR for any reason
- on advisory routes when the pilot wishes to use the advisory service
- using a Special Visual Flight Rules (SVFR) clearance if the destination aerodrome is to be notified
- when the Maximum Structural Take-Off Mass (MSTOM) exceeds 5700 kg and the flight is over 40 km
- crossing international boundaries.

Filing a flight plan is advised for flights:

- likely to go further than 10 nm from the coast
- over sparsely populated areas or areas where search and rescue operations would be difficult.

Booking out

It is a legal requirement that pilots intending to make a flight should inform the Air Traffic Service Unit (ATSU) at the departure aerodrome. Filing a flight plan makes booking-out unnecessary. On booking-out the information will not be transmitted to any other ATSU.

Flight plan submission times

The minimum time is 60 min before clearance to start up or taxi is requested. Exceptionally, operators should give as much notice as possible, but never less than 30 min. As already noted, if the flight is subject to flow management a minimum of 3 hours is required. Flight plans can be filed when airborne with any ATSU, but doing so may result in delays in the message being filed. If it is intended to enter Controlled Airspace (CAS) at least 10 min notice of entry must be given.

Persons on Board (POB)

This must be available to the ATSU for Search and Rescue Purposes (SAR) for the period up to the ETA at the destination plus 1 hour but, if this information has been sent to the operator's handling agency at the destination, no further action is required.

Action on diversion

If a pilot lands at an aerodrome other than the planned destination, the ATSU at the original destination must be informed within 30 min to avoid unnecessary action being taken by the alerting services.

Cancelling an IFR flight plan in flight

This is only permitted if the pilot is sure that without breaking any rules the flight may be continued under VFR.

Completing a flight plan

It is most important that the form is correctly completed down to the last detail. It will usually be handled by an automatic data processing system and computers are very intolerant of any variations from the standard format. Details of the procedure are given in a Yellow Aeronautical Information Circular (AIC) to which reference should be made. In a survey it was found that the following errors were common:

- omitting details of the route between points shown in item 15 (e.g. airways, advisory routes, direct (DCT) routes)
- not keeping to the laid-down formats (e.g. speed to always be four figures N0285 and not N285 for 285 knots)
- waypoints given not on the route selected
- FL requested not available on the route
- not using the approved ICAO aircraft designator (e.g. A320 and not EA32 for an Airbus 320)
- Standard Instrument Departures (SID) and Standard Terminal Arrival Routes (STAR) being included. They are not required.

Integrated Flight Plan Processing System (IFPS)

This is an essential part of the Central Flow Management (CFM) system as has already noted in this chapter. It is an important step in the European Civil Aviation Conference (ECAC) towards harmonised and integrated ATC systems within the 32 member states (see Chapter 1). All General Air Traffic (GAT), that is flights conducted under civil air traffic procedures, will be dealt with by the two IFPS units (IFPU). There is one in Belgium and one in France. Each has its own area of responsibility, but with a secondary role to support the other when the need arises. VFR plans will be dealt with by local ATC units.

Operators' flight planning procedure

Aircraft Operators (AO) are responsible for filing their IFR/GAT plans and submitting them to the IFPU and then getting the appropriate response:

- *ACK*
  plan accepted as submitted (a clearance).
- *MAN*
  plan not accepted and is awaiting manual interference by an IFPS operator. If this is successful, ACK will be sent together with the revised plan for checking.
- *REJ*
  plan rejected. Reasons will be given together with a copy of the plan received so that it can be checked for corruption. After correction a new plan will have to be filed as soon as possible.

Addressing flight plans

For flights wholly within the IFPS zone or entering it or overflying it, plans should be addressed to the two IFPUs. If using the AFTN there is a collective address (EGZYIFPS). Flights leaving the IFPS zone must also be addressed to the relevant ATSUs outside the zone.

Supplementary flight plan information

As an alternative to the ICAO procedure that the Supplementary Information stored in Field 19 should not be transmitted as it contains information that will only be required in the event of an emergency, the IFPS is able to accept and store this information for use if required.

Repetitive Flight Plans (RPL)

These apply to frequently recurring, regularly operated IFR flights with identical basic features. For flights within the IFPS zone, these will be filed with the CFMU at Eurocontrol. For flights involving routes that extend outside this zone, they will have to be filed with the national authorities of the affected external states which must have all agreed to accepting them. Types of submission:

- *New lists (NLST)*
  contains only new information (typically at the start of a new winter or summer season) and must be received by Eurocontrol a minimum of 14 days before the first flight.
- *Revised lists (RLST)*
  contains revised information particularly changes, cancellations or additional new flights. It must be received at least 7 working days (to include two Mondays) before the first flight affected by the amendment.

Specific Eurocontrol requirements for RPL operations

The basic principles for the submission are in ICAO documents 4444 and 7030. The following differences will be found:

- every ATS authority affected by the flight must agree to accept RPL. It can be taken that all states in the IFPS Zone accept the RPL
- to cancel or change an RPL the old RPL must be submitted in addition to the new one
- every NLST or RLST must be numbered sequentially starting at 001
- if an RPL is to be suspended it must be for at least 3 days and must be submitted at least 48 hours before the first affected flight.

Cancelling RPL for one day

To do this the AO sends an ICAO CNL (CANCEL) message to both the IFPS units, but not earlier than 20 hours before the EOBT. The same applies for a change (CHG) or delay (DLA) message. At 20 hours the RPL is sent to the the IFPS and becomes a standard Flight Plan (FPL).

Carriage of radio equipment – general (JAR-OPS 1.845)

A public transport operator shall ensure that the navigation and communications equipment required by JAR-OPS is:

- approved and installed as required and meets the minimum performance and the operational and airworthiness standards
- installed so that the failure of any required unit does not affect any other required unit

- in operable condition for the particular operation as provided in the Minimum Equipment List (MEL) (see Chapter 16)
- installed so that it is easily operable by any flight crew member needing to use it.

Radio equipment for VFR operations – JAR-OPS 1.860

An operator shall not operate an aeroplane under VFR over routes that can be navigated using visual landmarks unless it has the ability for the route being flown to:

- communicate with all necessary ground stations
- receive meteorological information
- reply to Secondary Surveillance Radar (SSR) interrogations.

Radio equipment for all other operations – JAR-OPS 1.865 on

- Two independent communication systems to communicate with all necessary ground stations for the route and possible diversions
- SSR equipment for the route
- VHF Omni-Range (VOR), Automatic Direction Finding (ADF) and Direction Measuring Equipment (DME)
- Instrument Landing System (ILS) or Microwave Landing System (MLS) as required for the flight and diversions
- An Area Navigation System (B-RNAV) if required for the route. It is now mandatory for flight on the entire ATS Route Network in the European Civil Aviation Conference (ECAC) if flying above a specified level. None of the 36 states are specifying a level below FL 100. The equipment will have to be to the RNP 5 standard which will ensure a track keeping accuracy of ±5 nm for 95% of the flight time
- An additional DME when some of the route is being navigated by DME alone
- An additional VOR when some of the route is being navigated by VOR alone
- An additional ADF when some of the route is being navigated by ADF alone
- If operating in Minimum Navigation Performance Specification (MNPS) airspace the ICAO specified equipment for this. For unrestricted operations this will be two independent Long Range Navigation Systems (LRNS) although on special routes this may be reduced to one

- If operating in airspace with Reduced Vertical Separation Minima (RVSM) two independent altimeters, an altitude alerting system, an automatic altitude control system and a SSR with altitude reporting system that can be connected to the altimeter in use.

Failure of radio navigation equipment

The standard ICAO procedure for this is described in the UKAIP En-Route Section (ENR):

- inform ATC and report altitude and approximate position
- ATC may help with navigation if radar is available
- ATC may authorise continuing flight into or in Controlled Airspace (CAS)
- if no authorisation, leave or avoid CAS and dense traffic areas and go to an area where flight may be continued in VMC or if this is not possible select a suitable area to descend through cloud and fly visually to a suitable aerodrome
- at all times keep ATC informed, take meteorological information and terrain clearance into account and make full use of Very High Frequency Direction Finding Stations.

Failure of two-way radio communications

Detailed procedures to be adopted are given in the aerodrome (AD) section of the UKAIP for each particular aerodrome. The general procedure is given in the En-Route Section (ENR):

- Continue the flight as planned to the holding point of the intended destination. Keep to the last assigned flight levels and then keep to the cruising levels planned. Operate the Secondary Radar Transponder (SSR) on Code 7600 with Mode C.
- If the transmitter might be operating, transmit position reports.
- Try to arrive over the holding point as near as possible to the last acknowledged ETA. If there is not one, the pilot should calculate from the last acknowledged position report and using the flight plan times. The standard inbound routes should be used.
- Begin to descend over the holding point at the last acknowledged Estimated Approach Time (EAT). If this does not exist, use the calculated ETA. The descent at 500 ft/min (fpm) must not be later than 10 min after the time it should have started. If ATC had stated 'Delay not determined' and so could not give an EAT, no attempt to land should be made, so use the following procedure.

- Land within 30 min of the time descent should have started. If this is not possible, but visual landing is possible, then do so. If not, leave the vicinity of the aerodrome and any associated CAS at the specified altitude and route as given in the AD section. If no altitudes or routes are specified, fly at the last given altitude or the Minimum Sector Altitude (MSA), whichever is the higher and avoid areas of dense traffic and then fly to an area which is VMC and land at a suitable aerodrome or select a suitable area, descend through cloud and fly to a suitable aerodrome and land as soon as possible. Inform ATC after landing.
- Similar procedures may be followed in the event of a missed approach.

Failure of SSR responder

If the transponder fails before departure and cannot be repaired:

- go to the nearest aerodrome where it can be repaired
- inform ATS, preferably before submission of the flight plan. Clearance to proceed may then be granted subject to certain conditions
- insert the letter 'N' in item 10 of the flight plan for complete unserviceability or, for partial failure, enter the appropriate letter indicating the transponder capability remaining serviceable.

If the transponder fails in flight, ATC will try to continue the planned flight or the aircraft will be ordered to return to the departure aerodrome or land at another mutually agreed aerodrome for repairs to be carried out. If this is not possible, the previous provisions apply.

Chapter 8
Flight at Aerodromes

Introduction

The ICAO references for this are Annex 14 and Procedures for Air Navigation Services (PANS) document 4444 (Rules of the Air and Air Traffic Services). For full details of aerodrome signals refer to the Visual Aids Handbook (CAP 637).

An aerodrome

Any area of land or water designed, set apart, equipped or commonly used to afford facilities for the take-off and landing of aircraft and includes any area or space wherever it may be designed, equipped or set apart for affording facilities for the landing and departure of aircraft able to descend or climb vertically.

Aerodrome Traffic Zones (ATZ)

These have the following dimensions:

- *Offshore installations (e.g. on oil platforms)*
  from mean sea level (msl) to 2000 ft amsl and within a radius of 1.5 nautical miles (nm).
- *Other aerodromes where longest runway is 1850 m (approximately 1 nm) or less*
  from the surface to 2000 ft above aerodrome level (aal) and within 2 nm of the mid-point of the longest runway. If this is not at least 1.5 nm beyond the end of any runway the following rule applies.
- *Aerodromes where longest runway is over 1850 m*
  as above but the distance from the mid-point of the longest runway is 2.5 nm.

ATZ are not given an airspace classification, but adopt the class of the airspace they are in.

Times of operation

These are as follows:

- *A government aerodrome*
  at the notified times as in the AD section of the UKAIP.
- *An aerodrome having an Air Traffic Control (ATC) or a flight information unit*
  during the notified hours of watch of the ATCU or the flight information unit.
- *A licensed aerodrome having two-way communication radio*
  during the notified hours of watch of the radio station.

During these times an aircraft cannot fly, take-off or land within the ATZ unless the pilot has obtained permission from the ATCU or the flight information unit. If neither of these are available, the pilot should obtain information from the ground/air radio station to enable a safe flight to be made in the ATZ.

Procedure when flying in ATZ

The pilot must listen out continuously and, if the aircraft has no radio, watch for ground signals. If the aircraft has a radio, position and height must be notified on entering and when about to leave the ATZ.

Procedure when flying in a Military Air Traffic Zone (MATZ)

Figure 2.4 in Chapter 2 describes the dimensions of a MATZ. A military airfield still has an ATZ as described above, but in addition, to provide increased protection, there may be a MATZ. Most of these have a penetration scheme for civil aircraft which is available during the normal hours of watch of the ATCU. This is as follows:

- call the controlling aerodrome when 15 nm or 5 min flying time from the aerodrome whichever is the greater
- when told to go ahead, pass all relevant information including intended destination
- comply with all instructions, keep a listening watch and inform when clear of the MATZ
- controller will provide radar separation, if possible. Otherwise a vertical separation of at least 500 ft will be used

- the normal altimeter setting used will be QFE but there are some exceptions. If there are a group of aerodromes, a 'clutch QFE' will be used. This is the QFE of the higher or highest aerodrome in the group.

A MATZ will be available during the published hours of watch but, outside these hours, pilots should call at least twice and then always proceed with caution. It should be remembered that because MATZ rules are not mandatory other aircraft may not be known to the controller. A yellow AIC refers.

Aerodrome signals (see CAP 637)

At aerodromes where general aviation movements are significant, visual aids displayed in a signals area 12 m square and bounded by a white border will be located so that it is easily visible from all directions of approach. The following signals may be seen in the signals area:

- take-off and landing to be made parallel to the shaft and towards the cross arm

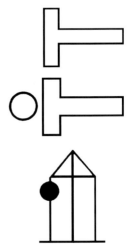

- take-off and landing directions do not necessarily coincide. Also indicated by a black ball suspended from a mast

- prohibition of landing (yellow on red)

- state of manoeuvring area is poor – take care when landing (yellow on red)

- ground movements confined to hard surfaces

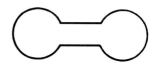

- take-off and land on runways. Ground movements not confined to hard surfaces

- the red L indicates that light aircraft may use the area on the airfield marked with a large white L

- arrows placed around two sides of the square indicates a right-hand circuit is in use. Coloured yellow and red. It may also be shown by a green flag

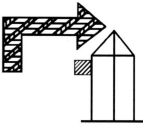

- glider flying is in progress. Two red balls suspended from a mast have the same meaning. A yellow cross on the airfield indicates the tow rope dropping area

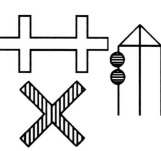

- helicopters must take-off and land in the area marked with a large white H

The following signals may be seen on or near to the control tower:

- visiting pilots should report here. The black C is on a yellow background

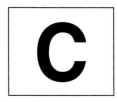

- aircraft must taxi only as directed by ATC. Alternatively a similar red and yellow flag may be flown

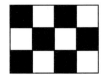

- indicates the runway in use. It may also have a R or L where there are parallel runways. The background is yellow

The following indicators may be seen in and around the airfield:

- a white cross indicates that the portion of the runway up to the next standard marking is unfit for use

- a yellow cross on a taxiway indicates that the portion of the taxiway up to the next standard marking is unfit for use

- areas considered unfit for use are marked along their boundaries by one of these orange and white markers

- if there is a displaced runway threshold these black and white markers are placed along both sides of the unusable runway area

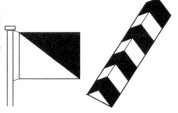

- an orange (or other conspicuous colour) windsock is placed near to the landing area where it is readily visible for all approaches. It may be lit at night.

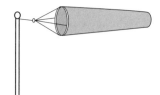

Visual signals from the control tower

Signal	Meaning to aircraft or vehicle on the ground	Meaning to aircraft in the air
Steady red light	Stop	Do not land. Give way and continue circling
Flashing red light	Move clear of landing area	Do not land. Aerodrome closed
Steady green light	Cleared to take-off	Cleared to land
Flashing green light	Cleared to move on the manoeuvring area	Return to aerodrome and await landing clearance
White flashes	Return to starting point on aerodrome	Land here on getting steady green
Red pyrotechnic or flare		Do not land. Cancels previous permission

Visual signals from aircraft in the air

Signal from aircraft	Meaning
Steady or flashing green or green flare	Night – may I land? Day – may I land in a different direction?
Red flare	Immediate assistance required
White flare or irregular switching of navigation or landing lights	I am compelled to land

Aerodrome circuit pattern (see Figure 8.1)

To ensure a safe and orderly traffic flow, a standard anti-clockwise traffic pattern is used at most aerodromes. Exceptionally, a clockwise (right-hand circuit) may be used usually in order to avoid sensitive areas or obstructions. The figure illustrates the standard visual circuit.

Priority of landing

The general rule is 'first come, first land'. If two aircraft arrive at the same time the lower has the priority. This does not apply if:

• an aircraft overtakes or cuts in front of another aircraft
• the ATCU has given an order of priority

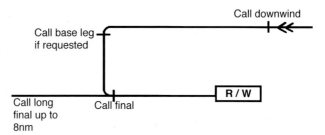

Figure 8.1 The standard visual circuit. Call Downwind when abeam upwind end of the run-way. Call Base Leg if requested by the ATC immediately on completion of the turn on to base leg. Call Final after completion of the turn on to final approach and when at a range of not more than 4 nm from the approach end of the runway. Call Long Final when making a straight-in approach, or a normal approach beyond 4 nm range, up to 8 nm. Again call Final when a range of 4 nm is reached.

• another aircraft wishes to land in an emergency. If this occurs the pilot who had to give way must obtain a fresh landing permission.

Use of runways

Normally only one aircraft should occupy the runway at a time, but an aircraft may land before the preceding aircraft has cleared the runway if:

• the runway is long enough for two aircraft
• it is daylight
• the preceding aircraft is clearly visible throughout the period
• the second aircraft has been warned by ATC saying 'land after ...'.

The ATCU will always nominate the runway to be used. Whether to use this runway will be the pilot's final decision.

Before take-off, if the instruction is received 'cleared to immediate take-off', the pilot should taxi on to the runway and commence take-off without stopping. If already on the runway, take-off without delay.

Closure of aerodromes

ATC may close a CAA or public-licensed aerodrome when the:

• landing area is unfit
• closure has been published in a NOTAM
• essential facilities have failed.

ATC has no authority to close an aerodrome for meteorological reasons. The arriving pilots will be in receipt of the latest weather information from either meteorological broadcasts (VOLMET) or from the Automatic Terminal Information Service (ATIS) broadcasts and will know the Aerodrome Operating Minima (AOM) that apply to their operation (see Chapter 18) and will have to decide whether to proceed with the landing or not. In bad weather conditions the first aircraft arriving will be told 'no delay expected'. If this aircraft holds to await an improvement, following aircraft will be told 'delay not determined' and will be put into the holding stack. Any of these may request permission to attempt an approach and the pilot will then be given descent and routing instructions and an Estimated Approach Time (EAT). In busy areas it is not usual for ATC to issue EAT when delays are likely to be under 20 min.

Aerodrome control service

This will be responsible for:

- giving departing aircraft all necessary instructions including permission to start engines, taxi, and take-off clearance. At large aerodromes these tasks may be handled by different controllers. It is quite common to have a ground movement controller
- arriving aircraft will be given their turn to land, runway in use, altimeter setting, wind velocity, details of any obstructions, taxi instructions after landing, etc. but, if there is a separate approach control, much of this information will be dealt with by the approach controller.

Typical departure procedure at a large airport

Forty minutes before the departure, the computer generates a departure strip giving a controller all the essential information about the flight. For regular flights a Repetitive Flight Plan (RPL) may have been filed some weeks or months previously and is now stored in the computer. When the passengers are on board, the pilot requests from the ground movement planning controller the Secondary Surveillance Radar (SSR) code (Squawk) to be used. When ready to start engines the pilot will be under the ground movement controller who will direct the aircraft to the runway. The air controller will then give the pilot take-off clearance. Immediately after take-off the pilot will be handed over to the radar departure controller or the Terminal Control Area (TMA) controller at the local Air Traffic Control Centre (ATCC). The procedure may well differ at various aerodromes, but the procedures described are fairly typical.

Approach control at aerodromes outside controlled airspace

Pilots flying outside the Aerodrome Traffic Zone (ATZ) are under no legal requirement to obey approach control instructions although they are strongly advised to do so when within 10 nm of the aerodrome and below 3000 ft if flying under Instrument Flight Rules (IFR). They are also advised to inform approach control 10 min before entry and put themselves voluntarily under their control. The approach control service operates as follows:

- departing aircraft cease to be under control when more than 10 min flying time away or when taken over by the local ATCC or when the pilot no longer wishes to be controlled
- transit aircraft remain under control until they are clear of the approach pattern or no longer wish to be controlled
- arriving aircraft contact approach control when released by ATCC or, if outside Controlled Airspace (CAS), when 10 min flying time away. The pilot will then be given:
  - runway in use
  - surface wind direction
  - visibility
  - present weather including significant cloud
  - altimeter setting information
  - other information including warnings (wind shear, gusts, runway surface state, etc.)
  - runway visual range.

Approach Control at aerodromes inside controlled airspace for IFR traffic

The procedure is in three stages (see Figures 8.2 and 8.3):

- *Initial approach*
  this takes place between the preceding en-route navigational fix and the facility to be used for making an instrument approach or a non-radio facility point associated with such a facility that is used to mark the end of the initial approach. This is the Initial Approach Fix (IAF).
- *Intermediate approach*
  between the IAF and the Final Approach Fix (FAF).
- *Final approach*
  between the FAF and the runway.

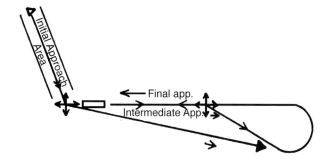

Figure 8.2 Plan view of approach stages.

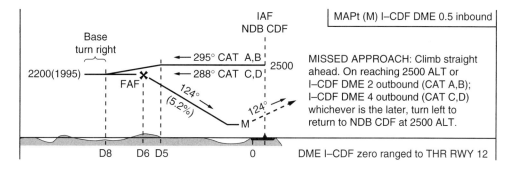

Figure 8.3 Side view of intermediate and final approach stages.

Holding (see Figure 8.4)

An aircraft may be required to 'hold' en-route, although in these days of good flow control this is much less common, or at the first facility at the destination aerodrome. With busy aerodromes and in poor weather conditions this is much more common. The holding procedure is as follows:

- *Basic pattern*
  this is a race-track pattern with 180° turns to the left or right as indicated on approach or area charts. Maximum speeds may also be specified.
- *Minimum holding altitude*
  this again will be shown on the charts. It will always give at least 1000 ft clearance over all obstacles.
- *Onward clearance time*
  when holding at a terminal facility, an Estimated Approach Time (EAT) will be given or possibly a 'delay not determined' message.

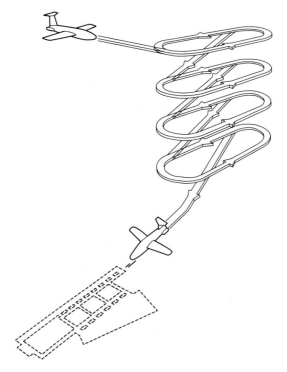

Figure 8.4 Holding 'stack' using a left-handed race track.

Diversion

This is flying from the intended destination to another aerodrome with the intention of landing there. Diversions will occur when:

- the weather is below the applicable operating minima
- the runway is obstructed
- an essential ground aid has failed
- the anticipated landing delay is unacceptable (probably because of the fuel state).

Diversions may be originated by:

- the pilot
- the operator of the aircraft
- exceptionally by the ATC.

When passing a diversion message to the pilot the ATC will:

- specify its origin

- give the reasons for the diversion
- if requested, pass the wind and weather data for the new route.

If pilots cannot comply they should say why and indicate their intentions.

Aerodrome Flight Information Service (AFIS)

This will be found at aerodromes where no approach or aerodrome control service is provided. It operates under a licensed AFIS officer (AFISO). Its purpose is to provide information useful for the safe and efficient conduct of flights within the Aerodrome Traffic Zone (ATZ). From this, the pilot can then decide the appropriate action to take. Pilots will always be aware that they are dealing with an AFISO because messages will include the word INFORMATION instead of 'Tower' or 'Approach' after the airfield name. An AFISO is responsible for:

- information to aircraft in the ATZ to assist the pilot to avoid collisions
- if danger threatens, giving information for avoiding other aircraft on the apron or manoeuvring area
- reporting on the state of the aerodrome and its facilities
- alerting emergency services
- initiating overdue action.

Aerodromes with air/ground communication service only

Here, the radio is operated by unqualified people and they should not provide any of the services offered by an AFISO. Their only function is to take and pass messages. They should always be identified in messages by the use of the word RADIO after the aerodrome name.

Rules for aircraft approaching aerodromes with an ATCU under VFR

In these cases the aircraft should not enter the ATZ without ATC permission except to observe the signals with a view to landing there. While in the ATZ the pilot should keep a continuous listening watch on the radio until taxiing is complete. All movements within the ATZ must be with ATC permission.

Rules for aircraft approaching aerodromes without an ATCU under VFR

Aircraft may enter the ATZ clear of cloud and at least 500 ft above aerodrome level (aal), for the purpose of observing the signals with view to landing there. Alternatively entry may be made with the permission of the person in charge of the aerodrome. On arrival:

- *Circuit procedure*
  if other aircraft are in the circuit, join the circuit by conforming with the traffic pattern.
- *Take-off and landing*
  use the directions shown by the ground signals. If no signals, land and take-off into wind unless good aviation practice demands otherwise.

If take-off and landing is not on the runway:

- *When landing*
  leave clear on the left any other aircraft ahead.
- *After landing*
  turn left after checking that it is clear to do so and then taxi clear of the landing area.
- *For take-off*
  leave clear on the left any aircraft ahead.

Rules for aircraft under VFR arriving at aerodromes with approach control

- *Initial contact*
  to be made when at the greater of 15 nm or 5 min flying time from the ATZ.
- *Frequency to use*
  if without the approach frequency use the aerodrome control frequency.
- *Information passed by ATC*
  landing information and known traffic so pilot can maintain separation from all other flights. If radar sequencing of other flights is in progress, VFR flights will be given information so that they can fit into the landing sequence.

Aerodromes: public transport of passengers and instruction in flying – ANO 90

For the aircraft covered by this article, but not including those operating under a police AOC, take-offs and landings are only permitted at:

- a government aerodrome or a CAA owned and managed aerodrome which is notified as available for such purposes or where permission has been given
- an aerodrome licensed for these purposes.

This article applies to:

- aeroplanes of over 2730 kg MSTOM engaged in:
  - public transport of passengers
  - flying instruction or tests to qualify for a licence or rating
- aeroplanes not over MSTOM of 2730 kg engaged in:
  - scheduled journeys for public transport of passengers
  - public transport of passengers beginning and ending at the same aerodrome
  - flying instruction or tests to qualify for a licence or rating
  - public transport of passengers on night flights
- helicopters and gyroplanes engaged in:
  - scheduled journeys for public transport of passengers
  - flying instruction or tests to qualify for a licence or rating
- gliders engaged in:
  - public transport of passengers
  - flying instruction, other than gliders flown under arrangements made by a flying club and only carrying club members
- the person in charge of an area, other than a properly licensed aerodrome, to be used by helicopters operating at night for the public transport of passengers shall ensure that lighting is installed to permit:
  - for landing, the landing area to be identified to the pilot, the landing direction to be shown and to ensure a safe landing can be made
  - a safe take-off to be made.

Range of TOWER, AFIS and air/ground communications

To minimise interference, communications should be kept to a minimum and restricted, as far as is possible, to heights up to 1000 ft aal.

Wake turbulence

The problems caused by wake turbulence (the disturbance of the air caused, particularly by large aircraft) is dealt with in a Pink Aeronautical Information Circular (AIC) which should be referred to. It gives the separations imposed by ATC to prevent wake turbulence problems.

Aircraft are categorised as follows to decide the amount of turbulence they will cause:

Category ICAO and flight plan UK: Heavy (H) 136 tonnes upward, 136 tonnes upward; Medium (M) 7–36 tonnes, 40–136 tonnes; Small (S) Not used, 17–40 tonnes; Light (L) Under 7 tonnes, Under 17 tonnes.

Aerodrome Reference Point (ARP)

This is a nominated point somewhere near the centre of the aerodrome from which the magnetic bearings and distances of obstructions within 4 nm that may constitute a hazard will be measured.

Details of these obstructions will be found in the Aerodrome (AD) section of the UKAIP and temporary hazards will also be found in NOTAMs.

Chapter 9
Flight in Other Types of Airspace

Introduction

The rules of the air regulations impose the following:

- *Rule 21*
  apart from gliders, all flights in class A airspace must be conducted under IFR unless a Special VFR (SVFR) clearance has been obtained.
- *Rule 31*
  any flight under IFR in any Controlled Airspace (CAS) requires a flight plan to be filed and a clearance obtained from an ATCU.
- *Rule 32*
  on any flight under IFR which is going to enter CAS, position reports, including the aircraft level, must be filed as required by the ATCU.

Controlling authorities of Control Zones

The designation of the controlling authority of a Control Zone (CTR) depends on the size and complexity of the zone. At smaller zones, the approach control unit is responsible. In larger zones there may be a zone controller and in the largest zones the Air Traffic Control Centre (ATCC) will usually be responsible.

Permission to enter controlled airspace (CAS)

This should be requested 10 min before entry time in the standard form:

- *Callsign or aircraft identification giving:*
  - aircraft type
  - position, level, altitude and flight conditions
  - ETA for point of entry
  - destination
  - true airspeed (TAS).

If the CAS boundary is less than 10 min flying time from the Flight Information Region (FIR) boundary, make the request as soon as possible. If the departure aerodrome is less than 10 min away, obtain clearance before take-off.

Pilots flying VFR wishing to enter CAS, when it appears IFR will be required, must contact the zone controller giving the time of arrival and await permission to enter. Entry into class A airspace, where IFR is required, is by prior clearance only unless it is being made in accordance with special local flying or entry/exit lane procedures. These procedures may give certain concessions from compliance with IFR for traffic using aerodromes within the CAS.

Flights on airways – R of A 21

All airways in the UK are class A airspace and so have to be conducted under Rules 21, 31 and 32 (see the first paragraph). Clearances must be obtained and the pilot must have an instrument rating (see Chapter 17).

Terrain clearance on airways

The following applies to all UK airways:

- *Lowest limit of airway*
  where the lower limit of an airway is defined as a Flight Level (FL), the absolute minimum altitude of the airway base will always be 1000 ft above the highest obstacle within 15 nm of the centre line.
- *Lowest usable FL*
  this will always be 500 ft above the base so providing a clearance of 1500 ft above the highest obstacle within 15 nm of the centre line.
- *Sections adjacent to CTRs*
  the lower limit will be not less than 700 ft above ground level (agl) and the lowest FL used will still be 500 ft above this.

Flight plan and clearance for public transport flight – JAR-OPS 1.300

A flight plan must be filed and a clearance obtained. Clearance will be in one of the following forms:

- *Clearance expires at a certain time*
  if not airborne by this time a new clearance is required.

- *Take-off not before a certain time*
  the pilot will have an idea as to when to request permission to start engines.
- *Unable to clear at FLs requested*
  ATC will offer an alternative. Accepting this will avoid or reduce delay.
- *Join airways at a particular time and place and not before a particular time*
  this will be used when the aircraft is departing from an aerodrome outside CAS.

Airborne procedures for airway flight

- *Joining or leaving positions*
  when cleared to join or leave at a particular point, the aircraft should be flown so as to cross the actual boundary of the airway as near as possible to the position. This indicates entering or leaving at as near as possible to a right angle with the centre line.
- *Visual Meteorological Condition (VMC) climbs or descents*
  aircraft on airways will be using IFR procedures in all weather conditions. However, when radar is not available ATC may offer VMC climb or descent clearances to avoid traffic delays. These will only be offered in the day, in VMC and if the pilot agrees to be responsible for affecting his or her own separation from other traffic. The ATC will provide information on other traffic in the vicinity.
- *Maintaining centre line*
  as far as possible traffic will be expected to operate along the defined centre line.

En-route holding on airways

As referred to in the previous chapter, en-route holding may be required to achieve the necessary traffic separation. Normally the tracks used will be parallel to the centre line of the airway and normally the turns will be to the right. The pilots will usually be given a specific time at which to leave the holding pattern. Pilots are required to report as follows:

- time and level of reaching the holding point
- time of leaving the holding pattern
- when leaving an assigned level for a new one.

Flights joining airways

Flight plans must be filed before departure or when airborne. Request joining clearance at least 10 min from the intended joining point. If any part of the flight is subject to Air Traffic Flow Management (ATFM) clearance must have been obtained from the appropriate Air Traffic Flow Management Unit (see Chapter 7). Joining clearance requires the following message to be passed:

- identification and aircraft type
- position and heading
- level and flight conditions
- departure aerodrome and ETA at entry point
- route and destination
- TAS and desired FL on the airway (if different to the one above).

Flights crossing airways in IFR

When the base of an airway is defined as a FL, an aircraft may without clearance cross the base at right angles. Pilots wishing to fly through an airway must have a valid Instrument Rating (IR), file a flight plan and will need to obtain crossing clearance at least 10 min from the intended crossing point.

When asked by ATC the following details will be required:

- identification and aircraft type
- position and heading
- level and flight conditions
- crossing position
- desired crossing level and estimated time of crossing.

Normally, the crossing clearance will be obtained from the Flight Information Region (FIR) controller and the aircraft should remain on this frequency throughout the operation. The controller will make a brief report on reaching the boundary giving identification, the airway, position, time and FL. Crossing should normally be at right angles to the airway and the aircraft should be in level flight before entering the airway.

Airways crossings or penetrations into VMC by civil aircraft

As already noted, crossings may be made, without clearance, at right angles across the base when the base is expressed as a FL. Other flights in VMC for special purposes (for example, photographic survey flights) may be permitted provided that:

- prior arrangements are made with the appropriate Area Control Centre (ACC)
- specific ATC clearance is obtained for each flight
- the aircraft can communicate on the appropriate airways frequency.

Advisory routes (ADR) description

These are not Controlled Airspace (CAS) and are in ICAO class F and so all the rules for flights in uncontrolled airspace apply. For example, in the UK the quadrantal flight level separation rules will apply. Advisory routes only exist for pilots wishing to avail themselves of the advisory service. Other traffic in the area may not be using the service and so their presence may not be known to the controlling authority. ADRs are 10 nm wide and are distinguished by a designator which includes a D; W6D, for example. Generally, the FLs used will give the same clearances as on airways (1500 ft above obstacles within 15 nm) but this cannot be guaranteed. Below 3000 ft above mean sea level, regional QNH should be used (see Chapter 4).

To use the service:

- IFR flight plans must be filed
- the lowest level applicable is shown on radio facility charts and other levels (in the UK) will follow the quadrantal rules up to the published maximum level (if any)
- an Instrument Rating (IR) is not required but if any part of the route passes through CAS an IR will be required
- when flight originates outside CAS or advisory airspace, the entry is by prior permission only. Flight plans must be filed at least 10 min before entry
- the radio and navigation equipment requirements will be that for any CAS that is associated with the advisory route (see Chapter 7)
- a listening watch must be maintained throughout the flight
- position reports must be sent at all designated reporting points. These will be marked by triangles.

Flights crossing ADRs

Pilots are advised to:

- contact the appropriate controller 10 min before the intended crossing and remain in contact until the crossing is complete
- if possible, crossing points should be associated with a radio facility and made at right angles to the ADR route
- pilots unable to make contact should cross at right angles and at the appropriate quadrantal level. Do not cross published holding patterns.

Flights in upper airspace

The UK Upper Information Region (UIR) from FL 245 to FL 660 is an upper airspace control area with the ICAO classification B. It contains a network of upper ATS routes. The following rules apply to flights in this area:

- a flight plan must be filed
- ATC permission must be obtained before entry
- a continuous radio watch must be maintained
- the flight must be in accordance with ATC instructions
- standard altimeter setting of 1013 mb will be used
- unless Reduced Vertical Separation Minima (RVSM) rules apply (see Chapter 3) cruising levels will usually be according to the semi-circular rules
- upper ATS routes in the UK are 10 nm wide and their designation begins with a U, for example UB 29.

Glider operations in class B airspace

Twenty-four hours before the intended flight contact should be made to the Airspace Management Cell (AMC) and then 2 hours before launch the following details should be confirmed:

- area concerned
- upper limit (if known)
- expected launch time, time of entry into and duration of flight within class B airspace
- number of gliders and their callsigns
- name and contact telephone number.

Following this, the activity will be discussed, a clearance issued and a frequency given. If a clearance cannot be issued, the reasons will be given and a new period suggested.

In flight:

- the glider will establish communication on passing through FL 200, maintain a listening watch and report again passing FL 245 on the descent
- checks will be made by the controller half-hourly while the glider is above FL 245. If the pilot does not receive a radio check, attempts should be made to establish contact and, if not successful, the pilot should descend below FL 245 within 15 min
- the pilot is responsible for remaining in the designated area
- all gliders must be fitted with appropriate radio and navigational equipment. If any of these become unserviceable, the pilot must descend to leave the class B airspace
- within the class B airspace, the pilot should listen to all flight information service messages with proximity warnings of co-ordinated traffic through the area. Pilots will be responsible for maintaining their own flight separation from other traffic
- in the area, position reports should be made when at airway/upper ATS route reporting points.

Use of conditional routes (CDR)

These are upper ATS Routes (described in UKAIP ENR 3.2) that are only available under certain conditions. There are three categories:

- only available as notified in the AIP
- can only be planned 24 hours in advance during the hours notified in a Conditional Route Availability Message (CRAM)
- only made available by ATC at short notice.

CDRs may have more than one category. CDRs have been introduced into UK airspace in accord with the concept of Flexible Use of Airspace (FUA) adopted by the European Civil Aviation Conference (ECAC). They are transit areas where Operational Air Traffic (OAT) unable to comply with the rules of the air take place. The CDR are only available when the airspace is not being used for such activities. The routes are managed by the UK Airspace Management Cell (AMC).

Special Visual Flight Rules (SVFR)

This provides a clearance for a pilot to fly within a Control Zone (CTR) although he or she is unable to comply with Instrument Flight Rules (IFR). In exceptional circumstances it may be granted to aircraft with a Maximum Structural Take-off Mass (MSTOM) over 5700 kg and capable of IFR flight. There are special conditions:

- SVFR is only granted when traffic conditions permit and normal IFR flights will not be affected
- for airfields within the CTR, special entry and exit lanes may be set up and compliance with rules laid down for them will be accepted as compliance with an ATC clearance
- without prejudice to weather limitations given in the aerodrome (AD) section of the UKAIP, ATC will not issue SVFR clearances to fixed wing aircraft when the visibility is 1800 m or less and/or the cloud ceiling is less than 600 m
- when under SVFR, the pilot must comply with all ATC instructions and remain clear of cloud and in sight of the surface. ATC may impose height limitations
- a full flight plan is not required but brief details of the callsign, aircraft type and the pilot's intentions must be given. A full flight plan is needed if the destination aerodrome is to be notified of the flight
- requests for SVFR may be made when airborne. All requests should give the entry time and be made 5–10 min beforehand
- ATC will give separation between all SVFR flights and other aircraft under IFR. This will not include other aircraft flying in exit/entry lanes
- the pilot is responsible for avoiding Aerodrome Traffic Zones (ATZ) unless prior permission to enter them has been obtained
- under SVFR pilots do not have to observe the requirement to fly 1500 ft above obstacles (see Chapter 11) but must observe other low flying rules. Particularly flying over congested areas aircraft, other than helicopters, must fly so that in the event of engine failure it could land clear of the area safely. Helicopters, whether over congested areas or not, must be at such a height as would enable a safe landing to be made in the event of engine failure.

Action after communication failure when operating under SVFR

- In the event of two-way communication failure, operate transponder on Code 7600 and Mode C if the equipment is available.
- Transmit blind if there is a chance that the transmitter is still operating.

- If not yet in the CTR do not enter.
- If in the CTR and due to land in the CTR fly in accord with the clearance and land as soon as possible. When in the circuit, observe aerodrome signals.
- If transitting the CTR, continue the flight above the cleared altitude to leave the CTR by the most direct route considering weather limitations, obstacle clearance and areas of known heavy traffic.
- In all cases notify ATC as soon as possible after landing.

Flights in class G airspace

This is often referred to as the 'free' or 'open' Flight Information Region (FIR). ICAO Standards and Recommended Practices (SARPs) require only a Flight Information Service (FIS) and an alerting service to be provided. In the UK other services may be available:

- an Air Traffic Service (ATS) to flights arriving at, departing from and overflying aerodromes in class G airspace. These services include aerodrome, approach and approach radar controls (see Chapter 8)
- Radar Advisory Service (RAS) and Radar Information Service (RIS) – see Chapter 10
- Flight Information Service (FIS).

Flight Information Service (FIS)

This is provided at the Area Control Centre (ACC) through an area FIS controller. In addition to the normal FIS of:

- meteorological warnings
- meteorological conditions at destination and alternative
- meteorological reports, e.g. if VFR possible along the flight path
- airfield serviceability
- airfield facilities.

The FIS controller will also provide the following services:

- receiving requests for joining or crossing Controlled Airspace (CAS) or advisory routes
- passing ETA to destination aerodromes in special cases, such as diversions, or at particular locations when traffic conditions demand it. Normally, however, pilots should speak directly to the ATC at the destination aerodrome at least 10 min before ETA

- accepting airborne flight plans
- operating a very limited warning system of proximity hazards. The controller may warn of known traffic in the vicinity and warn of the possibility of dangerous proximity. As the controller will not have information on all the traffic in the class G airspace, his or her information should not be guaranteed as providing a complete picture of all proximity hazards.

Non-Standard Flights (NSF) in Controlled Airspace (CAS)

These are flights involving aerial tasks which do not follow published routes or notified procedures. Typical examples would be:

- a formation of civil aircraft
- photographic survey work
- aerial advertising.

Applications to carry out these flights should be made at least 21 days ahead through an ACC and will need to give:

- purpose of the flight
- area concerned and proposed tracks to be flown
- estimated duration
- operating heights
- aircraft types and registrations
- departure aerodrome
- planned date and time.

Agreed applications will be given an NSF reference number. This is only an approval in principle and prior clearance must be obtained from the appropriate ACC on the day. Sometimes tasks may have to be abandoned when the aircraft is in flight. In contact with the FIS and quoting the reference number, a new area may be arranged for the task.

Unusual Aerial Activities (UAA) outside Controlled Airspace (CAS)

These may be a hazard if other traffic is not aware that they are taking place. The Airspace Utilisation Service (AUS) and the Civil Aviation Authority (CAA) need prior notification so that they may co-ordinate and notify the event or for the authority to issue a permission or exemption under the Air Navigation Order (ANO) and the regulations. There is a

Civil Aviation Publication (CAP) 403 giving guidance on 'Flying Displays and Special Events'. Typical UAA:

- unusual concentrations of aircraft – rallies or fly-ins
- activities requiring a permission or exemption from the ANO – low flying near assemblies of people, dropping of articles or parachutists or balloon or kite flying
- air shows, displays and races
- activities requiring the establishment of a temporary ATC unit.

The notice required by the authority and the AUS varies:

- when a temporary aerodrome licence is required – 60 days
- when a temporary aerodrome licence is not required – 42 days
- when over 100 aircraft involved – at least 90 days
- when a temporary ATCU is required – 90 days.

Low-level Civil Aircraft Notification Procedures (CANP)

For training purposes military aircraft often operate below 2000 ft above ground level (agl) and frequently below 1000 ft agl. Civil aircraft may be at the same levels on authorised flights for crop spraying and aerial surveys. These activities take place in class G airspace where collision avoidance is essentially based on 'the see and be seen' principle. The greatest risks are likely to be below 1000 ft agl when the pilot may be engaged on activities which inhibit look-out or reduces aircraft manoeuvrability. The CANP is designed to collect this information about civil aircraft for the information of the military authorities. The tactical booking cell (TBC) referred to as ALFENS OPS is responsible for co-ordinating this information. The following activities, expected to last more than 20 min at a particular site, should be notified to the TBC at least 4 hours beforehand giving all relevant information:

- aerial crop spraying
- underslung aerial load lifting
- aerial photography
- aerial surveys/air surveillance.

A yellow Aeronautical Information Circular (AIC) gives further details.

Helicopter pipeline/powerline inspection procedures

These operate below 1000 ft agl and are not normally able to predict their movements with sufficient accuracy to use the CANP system. A similar procedure to the CANPs procedure is used, which is known as the Pipeline Inspection Notification System (PINS). At least 4 hours' notice should be given to ALFENS OPS (see previous section) who will distribute the information to military pilots that may be affected. The pipeline inspection aircraft are recommended to operate between 500 and 700 ft agl where they will be above and skylined to the majority of low flying military aircraft which operate below 500 ft agl. As an additional precaution, the pipeline aircraft should set their transponders to squawk 0036 and mode C while doing their inspections.

A yellow AIC gives further information.

Royal flights in fixed wing aircraft

In the UK to protect Royal flights in fixed wing aircraft, the airspace around the route is designated 'purple airspace' within which the ATC applies special procedures which must be obeyed by all aircraft during the periods set out in a Royal NOTAM (R NOTAM), which will be issued to all interested parties. A yellow AIC gives details of a freephone service which will give up-to-date details about purple airspace. The following points should be noted:

- if possible, the Royal flights in fixed wing aircraft will use the airways system and the appropriate portions of the permanent Controlled Airspace (CAS) will be designated as purple airspace
- if going outside the CAS, a purple airway usually 10 nm wide will be set up where required
- if no Control Zones (CTR) exist at the aerodromes being used, a purple CTR usually of 10 nm radius will be set up to the FL being used for the flight
- purple airspace lasts from 15 min before to 30 min after the expected arrival and departure times of the Royal flight
- on flights longer than an hour the purple airways may be sectioned
- all purple airspace is class A
- gliders should not fly in purple airspace.

Royal flights in helicopters

Purple airspace is not normally established for these. Instead a Royal Low Level Corridor (RLLC) will be set up by R NOTAM and:

- the RLLC will be marked by a series of check points about 30 nm apart
- the R NOTAM will give the ETD/ETAs at the check points
- civil pilots flying near the Royal route should keep a good look out and maintain an adequate separation
- the R NOTAM will give details of aerodromes from which information on the progress of the Royal helicopter may be obtained.

Chapter 10
Use of Radar in Air Traffic Services (ATS)

Introduction

In general, the UK follows the procedures for the use of radar in ATS which are given in ICAO Doc 4444 with the important difference that the radar service provided outside Controlled Airspace (CAS) will be either an advisory service or an information service.

Types of radar service in the UK

Type of airspace	Service	ATC action with unknown aircraft
Class A – always IFR. Class D – CAS below FL 245 in which all flights are subject to ATC authority.	Radar control service	Traffic information and avoiding action will not be given unless it appears that the radar echo is a lost aircraft or one with radio failure.
Class E – CAS in which VFR flight permitted without ATC clearance.	Radar control service	Traffic information passed if it does not affect radar sequencing of traffic or separation of IFR flights. Avoiding action will be given if requested but to limits decided by the radar controller or if it appears that the aircraft is lost or experiencing radio failure.
Class B – upper airspace control area	Radar control service	*Within the military mandatory radar service area* Procedures exist to ensure separation between aircraft operating under different ATS units. Generally information will only be given if the aircraft appears to be lost or suffering radio failure. *Outside the military mandatory service area* When possible, traffic information will be given. If necessary, avoiding action may be given.
Class F – advisory routes	Radar advisory or information service	Traffic information and advice on avoiding action. Traffic information but no avoiding action. Pilot is responsible for his or her own separation.

Radar control service

This includes:

* radar separation to all aircraft
* monitoring en-route and approach aircraft
* radar vectoring to position aircraft for approach
* assistance in distress
* assistance in crossing CAS
* proximity hazard warnings
* information on observed weather
* navigation assistance.

Radar Advisory Service (RAS)

In this the controller gives advice so as to maintain the separation between aircraft using the advisory service and also he will pass information on conflicting non-participating aircraft together with advice to avoid the confliction. Under RAS the following conditions apply:

* *The service is only provided to IFR flights*
* *Advice may require flight to be made in Instrument Meteorological Conditions (IMC)*
  if pilots are not qualified to do this, they should advise the controller.
* *Pilots are not legally required to accept the controller's advice*
  if pilots choose not to, they should inform the controller and assume responsibility for their future actions.
* *Pilots must inform the controller if they change heading or level*
  the controller will try to obtain a separation of at least 5 nm or 5000 ft from non-participating traffic unless the authority requires otherwise.
* *Pilots are responsible for terrain clearance*
  but there will be a level below which the service will not be provided.

Radar Information Service (RIS)

In this, the controller will give details of conflicting traffic, but will not give any avoiding action. This will be the responsibility of the pilot. The following conditions apply:

* the service may be requested in any flight or meteorological conditions
* updates will only be given if the pilot requests them or if the controller considers that a definite hazard still exists

- controllers may provide vectors, but they will not be provided to maintain separation
- the pilot must advise the controller before changing level, level band or route
- RIS may be offered when RAS is not possible
- requests to change to RAS will be accepted subject to the controller's workload. Prescribed separation will be applied as soon as possible
- for manoeuvring flights involving frequent changes of heading or flight level, RIS may be requested. Information on conflicting traffic will be passed with reference to cardinal points. Pilots must indicate the level band they wish to fly in and are responsible for selecting the area to operate in. The controller may assist with advice
- pilots are responsible for terrain clearance. The controller will not provide vectors below a certain level unless the authority says otherwise.

Limitations of radar services outside CAS

The services may well be limited. If a radar controller thinks he or she cannot maintain a full service, the pilot will be warned to take the limitations into account in general airmanship. Limitation of the service may occur when the:

- aircraft is close to the limits of the radar coverage
- aircraft is close to permanent echo areas or strong weather returns
- aircraft is operating in areas of dense traffic
- controller suspects the performance of the radar
- controller is providing the service using Secondary Surveillance Radar (SSR) data only.

Terrain clearance for radar services

The controllers will see that the levels given to IFR flights in receipt of a radar control service or of a RAS will have at least the following clearances (see Figure 10.1).

- within 30 nm of the radar antenna, but excluding the final and intermediate approach areas, a clearance of 1000 ft above any fixed obstacle closer than 5 nm to the aircraft or which is within an area 15 nm ahead of and 20° either side of the track. The authority may reduce these to 3 and 10 nm, respectively. Levels given for the initial approach will give the same clearances.

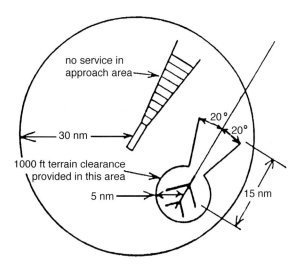

Figure 10.1 Areas to consider for terrain clearance.

- outside 30 nm from the radar antenna, for flights on airways or advisory routes, the clearance used will be 1000 ft above any fixed obstacle within 15 nm of the centre line. Otherwise 1000 ft above any fixed obstacle within 30 nm of the aircraft.

Radar controllers have no responsibility for the terrain clearance for aircraft under RIS or aircraft operating under Special VFR (SVFR) or VFR.
  A yellow AIC provides details of services to aircraft outside controlled airspace.

Other assistance provided by radar controllers

- Identified aircraft operating within Controlled Airspace (CAS) are deemed to be separated from unknown aircraft in adjoining uncontrolled airspace. The controller will aim to keep aircraft under control at least 2 nm within the CAS boundary.
- If on evidence of clutter on the radar screen the controller considers it sensible for the aircraft to leave CAS to avoid bad weather, the aircraft will be advised. The pilot may wish to use weather radar to detour the dangerous weather. If the pilot wishes to do so while remaining in CAS, clearance should be obtained from the radar controller. If the pilot has to leave CAS, it is necessary to request permission to re-join.

Methods of identification

Before any radar service can commence, the controller will need some positive identification of the aircraft by one of the following methods:

- pilot reporting over a facility
- at the controller's request, aircraft carries out a manoeuvre
- blip on the radar screen from a departing aircraft
- secondary surveillance radar (SSR) identification. Aircraft's squawk or flight number or identification may be displayed on the screen
- radar handover from one area controller to another
- pilot report from his radio fixing equipment (R Nav).

Secondary Surveillance Radar (SSR)

In this system a ground transmitter (interrogator) sends a series of coded pulses to which the aircraft equipment (transponder) replies. The reply will include a Mode A discrete four-figure identification number and, if Code C is being used, the FL of the aircraft. On the ground screen the squawk may be displayed or, by computer processing, it may be translated to the aircraft identification or the flight number. SSR is used by the controller to establish the aircraft identity and its FL so as to assist in the control of the aircraft. Generally primary radar is used to provide separation and control. SSR may be used to separate transponding aircraft as follows:

- in an emergency
- when primary radar fails and until non-radar separation is established provided the position and accuracy of the secondary information has been verified and the pilot has been advised.

Carriage of SSR

JAR-OPS 1.865 (b) just requires SSR to be carried when the route regulations call for it. In the UK a notification in the En-Route (ENR) section of the UKAIP requires SSR with mode A and mode C altitude reporting to be carried in the whole of UK airspace above FL 100 and below FL 100 if flying IFR. In the Scottish TMA, SSR is required from altitude 6000 ft to FL 100 in all conditions.

Exceptions to these rules:

- gliders
- aircraft below FL 100 receiving a crossing service through CAS.

Operation of transponders

Unless using special-purpose codes as described in the next section, pilots shall:

- if coming from an area where a code was allocated, maintain that code until told to change
- only select or reselect codes or switch off when airborne when told to do so
- acknowledge code setting instructions by reading back the code to be set
- set mode C simultaneously with mode A unless otherwise instructed
- when reporting levels under routine procedures or when requested, give the current altimeter reading to the nearest 100 ft. The controller can then verify the mode C data being transmitted. If the controller gets more than 200 ft difference, the pilot will normally be required to switch off mode C. If this cannot be done independently, the pilot will be asked to switch to mode A 0000 to indicate transponder malfunction.

Special-purpose SSR codes

Internationally the following mode A codes are reserved:

- *Code 7700*
  to indicate an emergency. This should be selected as soon as possible after the emergency is declared. If the aircraft is already using an allocated code and receiving an Air Traffic Service (ATS), this allocated code may be kept at the discretion of the controller or the pilot.
- *Code 7600*
  to indicate a radio failure.
- *Code 7500*
  to indicate unlawful interference (for example, hijacking) unless circumstances indicate the use of Code 7700.
- *Code 2000*
  when entering UK airspace from an airspace where no operation of transponders was required.
- *Code 7007*
  allocated to aircraft on airborne observation flights under the Treaty on Open Skies.
- *Code 0033*
  unless a discrete code has been given, to be used from 5 min before a parachute drop starts until the parachutists are estimated to have landed.

- *Code 0036*
  Helicopters on pipeline inspections.

Conspicuity code

This is 7000 which should be selected with care due to the proximity of some of the special-purpose codes. It should be used together with mode C when operating above FL 100 unless already using an allocated code or when using a special-purpose code.

Action in the event of transponder failure before departure

- Plan to proceed as directly as possible to the nearest suitable aerodrome where a repair can be made.
- Inform ATS as soon as possible, preferably before submitting flight plan. ATS may well, considering all factors, vary the flight plan.
- Insert in item 10 of the flight plan under SSR, the letter N for complete failure of the SSR or a letter corresponding to the remaining SSR capability.

Lower Airspace Radar Service (LARS) availability

This is available to all aircraft flying outside Controlled Airspace (CAS) up to FL 95 and within the limits of the radar/radio cover which is approximately 30 nm. Unless the ATS unit is operational for 24 hours (H24), the service will only be available during the working hours of the ATS unit.

Description of LARS

- The service provided will be a Radar Advisory or Information Service (RAS or RIS).
- Requests for an upgrade from RIS to RAS will be accepted if practicable. If RAS is not available, the controller will continue to offer RIS.
- Outside regulated airspace the radar service may be limited. See limitations of radar service outside CAS earlier in this chapter.
- In an emergency, pilots will be given all possible assistance.

Procedure for using LARS

- When approximately 40 nm from a participating unit, establish radio contact.

- When requested, provide details of the flight and state the service required.
- Maintain a listening watch.
- Follow the controller's advice or indicate why this is not possible.
- Advise when the service is no longer required.
- A separation of 3 nm between two identified aircraft working will be aimed for. Between an identified aircraft and another observed aircraft, 5 nm will be used unless a vertical separation of 1000 ft is known to exist.
- When possible aircraft will be handed over from one LARS unit to another overlapping one.
- If pilots wish to enter CAS, they are responsible for obtaining the necessary clearance although the LARS controller may be able to assist.
- Terrain clearance is the pilot's responsibility. The LARS unit will set a limit below which RAS will not be provided.

Military middle airspace radar advisory service

This provides a service similar to LARS to all aircraft operating outside regulated airspace and above FL 100. In most areas it extends to FL 240. Aircraft wishing to use the service must be fitted with transponders.

Approach radar services

This operates at selected aerodromes and provides services similar to the radar control service and the Radar Advisory Service (RAS). In addition the following may be provided:

- vectoring and sequencing of terminal traffic for standard approach aids
- monitoring Instrument Landing System (ILS) approaches in certain weather
- surveillance radar approaches (SRA). This has no height-finding capability and so no glide path information can be given but check heights will be given
- precision approach radar (PAR) mostly used by the military. In this both heading and glide path instructions will be given, often to within 400 m of the end of the runway
- weather information will be given as soon as possible after the aircraft has been accepted by the controller
- if during the approach the radar has weather clutter, the pilot will be asked to monitor the approach by other means. If these are not available, the pilot will be given overshoot instructions

- in the intermediate phase of the procedure, the pilot will be asked both Aerodrome Operational Minima (AOM) and the Missed Approach Point (MAP) for the runway in use. Pilots will not be given the Obstacle Clearance Height (OCH) or the MAP unless they are requested.
- SRA procedures, in line with ICAO recommendations, generally permit descent on final approach to the OCH without regard to the advisory heights given by the controller. These advisory heights are only given to the pilots as a guide to maintaining a constant angle glide path
- a MAP is normally located where the radar approach finishes – the Radar Termination Range (RTR). The MAP for a 2 nm SRA may be given as 1 nm after the RTR that will be 1 nm from the threshold.

Chapter 11
Airspace Restrictions, AIRPROX Procedures, Low Level Rules

Introduction

This chapter deals with the hazards and dangers that an aircraft may encounter either as a result of entering areas dangerous to aviation or from ground obstacles or the proximity of other aircraft.

Danger areas

This is an area in which activities dangerous to aircraft may exist at given times. The following are typical activities that may be dangerous to aircraft:

- military training areas
- navy firing ranges
- flights by pilotless aircraft
- bombing ranges
- rocket firing.

These areas are described in detail in the En-Route (ENR) section of the UKAIP together with a map showing all the areas and details of when they are active. They are of two types:

- scheduled danger areas which are enclosed by solid red lines on the map. These areas are always active at the stated times
- notified danger areas which are only active when published in NOTAMs and are enclosed by a pecked red line.

En-route charts show these areas together with details which may appear next to their location or, if the map is too crowded, in the margin of the map. Figure 11.1 shows a typical presentation.

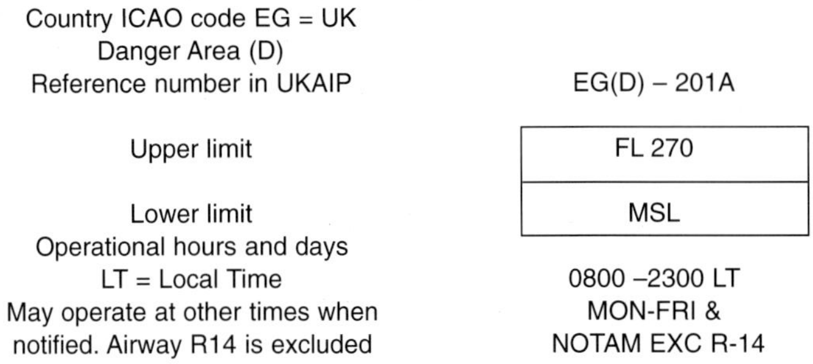

Country ICAO code EG = UK	
Danger Area (D)	
Reference number in UKAIP	EG(D) – 201A

Upper limit	FL 270
Lower limit	MSL

Operational hours and days
LT = Local Time
May operate at other times when
notified. Airway R14 is excluded

0800 –2300 LT
MON-FRI &
NOTAM EXC R-14

Figure 11.1 Danger area details on en-route chart.

Crossing and information services for danger areas

The UKAIP indicates danger areas which have available:

- a crossing service (DACS) which can offer crossing clearances
- an activity information service (DAAIS) – this will have no authority to offer a crossing clearance.

Restricted and prohibited areas

A prohibited area is one into which entry is absolutely prohibited at any time. It is indicated by the letter P and, on the chart of airspace restrictions, is surrounded by a solid purple line. A restricted area is one in which flight is restricted according to certain conditions and is also surrounded by a solid purple line. On en-route charts the only indication of these areas is by the use of the letters P and R.

Temporary danger and restricted areas

When considered necessary for the safety of life or property and particularly for those engaged in search and rescue activities, the appropriate Emergency Controlling Authority (ECA) may try to inhibit flight in the vicinity of an emergency incident. Depending on the nature of the incident, the initial action will normally be the introduction of a Temporary Danger Area (TDA) notified by NOTAM. There is no legal requirement for pilots to avoid these areas, although they are strongly advised to do so. If the TDA fails to produce the desired result, a Temporary Restricted Area (TRA) may be introduced making it an offence to enter the area without the permission of the ECA. Again this will be notified by NOTAM. Details of TRAs may also be given in the Mauve Aeronautical Information Circulars (AIC). A yellow AIC gives further information on the general subject.

Other potentially dangerous areas

These include the following which are all described in the UKAIP:

- military training areas which are usually in the upper airspace above FL 245
- areas of intense air activity (AIAA) – the UKAIP also has a special map of these
- aerial tactics areas (ATA) also shown on the above map together with low flying areas
- advisory radio areas
- air-to-air refuelling areas which are shown on the chart of airspace restrictions
- high intensity radio transmission areas also shown on the chart
- gas venting sites also shown on the chart
- small arms ranges.

Quite a number of these will be shown on the en-route charts.

Legal aspects relating to danger areas

A pink AIC indicates that there is no law prohibiting a pilot entering active Danger Areas (DA) although many DA have bye-laws restricting entry. It also emphasises that a pilot entering an active DA could be in breach of one of the following Articles of the Air Navigation Order:

- a commander of a UK registered aircraft should satisfy himself before take-off that the flight can be made safely
- nobody must act in a negligent or reckless manner likely to endanger an aircraft or anybody on board
- nobody must act so that an aircraft endangers life or property
- the commander of an aircraft flying in a DA must comply with all instructions given by the appropriate authority.

Danger of entering an apparently inactive danger area

Although a DA, notified as active, is apparently not so the following may still cause problems:

- *Weapon and missile firing*
  even when an infringing aircraft is detected, it may take some time to close down the activities.
- *High energy manoeuvres*
  aircraft performing these and possibly releasing weapons may be unable to comply with the rules of the air for collision avoidance (see Chapter 17)
- *Parachuting*
  the visual sighting of free-falling bodies is very difficult to see.

Legal aspects of restricted areas – ANO 75

The authority may restrict or prohibit flying over certain areas in the following circumstances and details will be issued in a Mauve Aeronautical Information Circular (AIC):

- when there is a large gathering of people expected
- in the vicinity of any aircraft event – contests, exhibitions, etc.
- for reason of national defence or any other reason affecting public interests.

If an aircraft contravenes the last of these items, the pilot must take the following action:

- not commence descent while over the area
- leave the area by the shortest possible route
- comply with any instructions by the controlling authority.

Air navigation obstructions

The UKAIP lists in the en-route (ENR) section all known obstructions in the UK which are more than 300 ft above ground level (agl). These will also be shown on aeronautical topographical maps. Any that are above 150 m (392 ft) will definitely be lit by night, but many below this may also be lit. The usual lighting is a white flashing strobe light but some may use red fixed or flashing lights. Many are lit during the day. Within 4 nm of an airfield any obstruction that is considered to be an obstruction to traffic using the airfield will be given in the aerodrome (AD) section of the UKAIP together with whether or not it is lit and also the location. This is often given as a true bearing and distance from the Aerodrome Reference Point (ARP).

Obstructions marked in the UKAIP as having a gas flare (FLR) are burning off high pressure gas. The flames may reach a considerable height and may not easily be visible in bright sunlight.

Other hazards

There are many other potential hazards listed in the ENR section of the UKAIP:

- *Glider launching sites*
  a particular hazard is the launching cable which is often dropped from 2000 ft agl and at some sites this may be exceeded.
- *Hang-gliding and parascending winch/auto-tow launching sites*
  the vertical limits of these is not over 2000 ft agl.
- *Hang gliding foot-launched sites*
  no vertical limits are quoted for these.
- *Free-fall drop zones*
  it is quite common for these to operate up to FL 150. Night operations may take place at any of the zones, but this is usually notified by NOTAM.
- *Microlight sites*
  no vertical limits are quoted for these.

Bird migration

There are no well-defined heavily used bird migratory routes in the UK at any time of the year. Bird migration to and from the UK occurs regularly largely in autumn and winter on broad fronts in streams running from east to west and from north to south. The density is greatest in south-east England where the two fronts cross. Most migrating birds fly below 5000 ft. Offshore islands, cliffs, inland waters and shallow estuaries attract flocks of birds at various times of the year. Within 20 nm of such locations concentrations of birds flying mostly below 1500 ft may be encountered. The UKAIP in the en-route (ENR) section gives maps showing these areas at different times of the year. It is sensible to avoid flying below 1500 ft in these areas. If flying below 1500 ft, remember that the risk of bird strike increases with speed. Few cases are reported of bird strikes on aircraft flying slower than 80 knots.

Bird sanctuaries

In the UK there are some bird sanctuaries which are shown on the airspace restrictions chart. Pilots are asked to avoid them, especially in the breeding season (given in the UKAIP ENR section) by keeping above the altitudes quoted (2000–4000 ft).

Release of racing pigeons

There will be no release of racing pigeons within 7 nm of 26 main airports in the UK. For other aerodromes, the Senior Air Traffic Control Officer (SATCO) should be notified in writing at least 14 days beforehand and then telephoned 30 min before the actual release time. Sometimes it may be necessary for the SATCO to delay the liberation.

AIRPROX

An AIRPROX report should be made whenever a pilot or controller considers that the distance between aircraft as well as their relative positions and speed have been such that the safety of the aircraft involved has been compromised. Where the event involves a UK registered public transport aircraft of over 2300 kg, it must be reported to comply with the Mandatory Occurrence Reporting (MOR) procedures (see Chapter 16).

AIRPROX procedures

AIRPROX reports made by pilots – AIRPROX (P) and those by controllers – AIRPROX (C) are dealt with separately. The procedure for pilots is as follows:

- an initial report should be made immediately by radio to the Air Traffic Service Unit (ATSU) with which the pilot is in communication, prefixing the message with AIRPROX
- the essential information should follow the sequence of items shaded on the AIRPROX (P) report forms which are available at all CAA ATSU
- if a radio report cannot be made, an immediate report should be made on landing to any ATSU but preferably to an Area Control Centre (ACC)
- the Aeronautical Fixed Telecommunication Network (AFTN) may be used from places abroad when the AIRPROX (P) could not be made by radio at the time.

Investigation of AIRPROX

The primary reason for investigating reports is to find the cause and so lead to action to reduce the possibility of future collisions. All concerned parties will be advised of the findings.

AIRPROX in foreign airspace

The procedures to be followed should be as laid down in the foreign state's AIP. The CAA has no authority to investigate these incidents, but is concerned about them, particularly when UK public transport aircraft of over 2300 kg are involved. Copies of the confirmatory reports made and any responses should be sent to the CAA.

Low-level flying over towns, cities and settlements

If flying along a route that is notified as such in the UKAIP or obeying ATC instruction when on Special VFR (SVFR), the lowest height that can be flown is that from which, in the case of the failure of an engine, the area can be cleared without causing danger to persons or property on the surface. On any other flight the lowest height is the greatest of:

- 1500 ft above the highest obstacle within 2000 ft of the aircraft
- that height from which the area can be cleared without damage to persons or property on the surface. If towing a banner such height that the banner shall not be dropped within the congested area

It has been held in court that flying over central London under a SVFR clearance which complied with the first of these, did not prevent the flight being illegal under the second proviso. Helicopters may be given permission by the CAA in writing to exempt them from the first of these conditions.

Low-level flying over assemblies of people

Unless it is a police aircraft or taking part in an air display, it is forbidden to fly over or within 3000 ft of any assembly of people in the open air of more than 1000 persons without the written permission of the CAA and the written consent of the organiser. The height must also make allowance for the possibility of engine failure under the same conditions as in the previous section.

If flights do not comply with these conditions, it is a recognised defence to show:

- the flight was being made at a reasonable height
- it was not connected with the assembly in any way.

Low-level flying over persons, vessels, vehicles and structures

If not engaged in take-off and landing, or a glider hill soaring, or picking up or dropping tow ropes, banners, etc. at an aerodrome, or a police aircraft, or taking part in an air display, flights must not fly closer than 500 ft to any person, vehicle, vessel or structure.

General exceptions

The following exceptions apply to all the above three categories of low flying:

- taking-off, landing and practising approaches at an appropriate place
- flying for the purpose of saving life
- captive balloons or kites.

Exceptions for aviation events

The rules forbidding flight within 500 ft to persons, vessels, vehicles or structures and not within 3000 ft of an assembly of over 1000 people can be waived, if the flight is made with the consent of the organisers, and the event consists:

- wholly or partly of an aircraft race or contest
- wholly or partly of a flying exhibition for which CAA permission has been obtained and the flights are made in accord with that permission
- of an event at which the attendance is likely to be less than 500 persons
- of an event at a Ministry of Defence aerodrome.

Chapter 12
Meteorology

Introduction

This chapter deals with the organisation and procedures adopted by the meteorology services for aviation in the UK. The UK follows the internationally agreed procedures and so similar procedures will be found in most other countries. The relevant ICAO documents are:

- Annex 3 – Meteorological Services for Air Navigation
- Doc 7754 – Air Navigation Plan for EUR region
- Doc 8896 – Manual of Aeronautical Meteorological Practice.

As required, the differences between these are notified in the GEN section of the UKAIP but they are very slight.

Organisation

- Meteorological services for civil aviation in the UK are provided by the National Meteorological Centre (NMC or Met Office) at Bracknell which acts as the Meteorological Watch Office (MWO) for the London and Scottish FIR/UIR and the Shanwick FIR/OCA. The main function of an MWO is to keep a continuous watch over the assigned area and to:
  - advise ATC on diversions
  - liase with other meteorological offices in the area
  - provide en-route forecast service
  - initiate SIGMET messages (see in flight procedures).
- Most major airports in the UK have meteorological offices, which are linked by a telecommunications network. In most cases they operate on a 24-hour basis. Routine reports (METARs) are made every 30 min, in most cases, and the information is distributed throughout the network. Special reports (SPECIs) are issued locally at ATS units between routine reports whenever a significant change occurs.

Collection of information

- It is important that the observation, from stations scattered over a wide area are collected as quickly as possible. These are collected by the NMC together with information from overseas and from satellite observations.
- As well as surface observations, a few selected stations supply information about high-level conditions. Radio sonde balloons are sent aloft twice daily. These carry a set of instruments which record and transmit details of temperature, pressure and humidity. Radar tracking will also record the position of the balloon enabling the calculation of high-level winds. These balloons go up to 35 km (115 000 ft).
- All the collected information is fed into a computer which assembles the data for transmission to airports and other users.

Pre-flight briefing

The main method is self-briefing using the information and documentation displayed in the briefing areas. Alternatively, flight crew and operators may obtain information direct by using the Met Office's PC service (MIST, Met Information Self-briefing Terminal), the fax services provided by the Met Office and CAA and the CAA's AIRMET telephone service.

AIRMET service

This is a general aviation briefing service. The basic service consists of ten routine forecasts in plain language and covering the UK and the near continent and a comprehensive selection of TAFs and METARs for the same areas. Information is given in spoken form via the public telephone network and in text form via the Aeronautical Fixed Telecommunication Network (AFTN), telex and facsimile. The forecasts will reflect the contents of SIGMETs that are current at the time of issue or amendments of the forecasts. Safety-related amplification may be obtained from one of the forecast offices but callers must confirm that they already have the current AIRMET forecast.

Special forecasts

When the standard self-briefing is insufficient, a special forecast may be obtained from the last UK departure point to the first transit airport outside the standard coverage. This will be supplied by the appropriate forecast office for the departure point. By prior arrangement, forecasts for further legs can be obtained, provided the flights are under 6 hours and no stop-overs over 1 hour are planned. These special forecasts will not be provided for flights wholly within the AIRMET coverage.

Notification times required for special forecasts

For flights up to 500 nm, 2 hours before collection time and for flights over 500 nm, 4 hours before collection time.

Observations and reports

- *Climatological information*
  this is available for certain aerodromes according to the following categories:

  A readily available, based on at least 10 years of at least 3-hourly data.
  B readily available, based on less than 10 years and/or some night-time gaps
  C limited, based on METAR reports since 1983 or later
  D insufficient data to provide reliable statistics.

- *Observations at aerodromes*
  most UK aerodromes make meteorological observations and reports, some only during the aerodrome operational hours and some for 24 hours (H24). In the majority of cases the reports are made half-hourly but some are only hourly. The following types of reports may be made:
  – actual weather (METAR)
  – meteorological synoptic reports (SYNOP)
  – Aerodrome Warnings (AW)
  – Marked Temperature Inversions (MTI)
  – Runway Visual Range (RVR)
  – Climatological Data (see above).

Aerodrome warnings

These are issued when one or more of the following occurs or is expected:

- gales (mean surface wind above 33 knots or if gusts expected to exceed 42 knots)
- squalls, hail or thunderstorms

- snow
- warnings are issued when any of the following are expected:
  - ground frost
  - air frost
  - hoar frost, rime or glazed ice deposited on parked aircraft
  - fog (usually with visibility below 600 m)
  - rising dust or sand
  - freezing precipitation
  - Marked Temperature Inversion (MTI) at some aerodromes when a temperature difference of +10°C exists between the surface and any point up to 1000 ft above the aerodrome.

Aerodrome operators requiring these warnings should apply to the Met authority and they will usually be sent by Aeronautical Fixed Telecommunication Network (AFTN), telephone message or fax. The aerodrome operator is then responsible for passing on the information.

Broadcast meteorological charts

These are distributed by two facsimile services. Broadcast fax is a service provided for users needing a regular supply of charts each week. MET-FAX is a dial-up service used mainly by general aviation for requesting charts as required for the UK and continental Europe. Charts routinely sent on the broadcast fax network cover:

- low- and medium-level flights over the UK and nearby continent
- medium- and high-level flights to Europe and the Mediterranean
- high-level flights to North America
- high-level flights to the Middle and Far East
- high-level flights to Africa.

Other charts are available on request.

Broadcast text meteorological information

Aerodrome Meteorological Reports (METAR), Aerodrome Forecasts (TAF) and warnings of weather significant to flight safety (SIGMET) including volcanic activity reports are broadcast by teleprinter throughout the UK and internationally. The following may be added to the METARs:

- *TREND*
  short-term landing forecast valid for 2 hours.

- *Runway state*
  added when weather conditions require and continues until they have ceased.

In addition, Special Aerodrome Meteorological Reports (SPECI) are issued locally when conditions change beyond certain limits.

Flight procedures – SIGMET messages

Meteorological Watch Offices (MWO) are responsible for preparing and distributing SIGMET messages to all Area Control Centres (ACC) and Flight Information Centres (FIC). These are messages concerning en-route weather phenomena that may affect the safety of aircraft operations. The ACC/FIC should warn aircraft of any of the following SIGMET phenomena for the route ahead for up to 500 nm or 2 hours flying time.

- thunderstorms
- heavy hail
- tropical cyclones
- freezing rain
- severe turbulence or severe icing (not associated with convective cloud)
- severe mountain waves
- heavy sand/dust storms
- volcanic ash clouds.

Flight procedures – VOLMET

Regular continuous broadcasts of weather reports and tendencies are made by Air Traffic Control Centres (ATCC). These VOLMET broadcasts cover selected aerodromes. Each cycle is preceded by a time announcement indicating the end of the reporting period. Individual aerodrome reports are broadcast for 30 min after the observations and then a further 30 min if no fresh observations are available. After that, the aerodrome's reports are suspended. The reports consist of:

- surface wind
- visibility (or CAVOK) – see below
- RVR (if applicable)
- weather
- cloud (or CAVOK)
- temperature

- dewpoint
- QNH (mean sea-level pressure)
- recent weather, if applicable
- windshear, if applicable
- TREND, if applicable – see below
- runway contamination warning, if applicable
- SNOCLO, if applicable. Indicating that the aerodrome is closed by snow.

CAVOK (Cloud and Visibility OK) is used when:

- visibility is 10 km or more
- no cloud below 5000 ft or below the minimum sector altitude if greater and no cumulonimbus cloud
- weather – no precipitation or thunderstorms, shallow fog or low drifting snow.

TREND, when included, indicates the trend of the weather during the next 2 hours. It follows the main body of the broadcast, with one of the following indicators:

- *TREND*
  trend forecast.
- *GRADU*
  gradual change at a constant rate.
- *RAPID*
  change in 30 min or less.
- *INTER*
  intermittent change – conditions varying.
- *TEMPO*
  change likely to last less than an hour.
- *NOSIG*
  no significant change expected.
- *PROB*
  percentage probability.

Further in-flight procedures

In-flight information to aircraft in flight is supplied in accordance with area meteorological watch procedures supplemented by an en-route forecast service. Information is also available from the appropriate ATSU at the commander's request. An in-flight en-route service is available in exceptional cases by prior arrangement, details of which will be found in the GEN section of the UKAIP. A meteorological office will be selected to provide the aircraft in flight with the winds and temperatures for specific route sectors. Applications will have to be made in advance stating:

- the flight levels and route sectors required
- the period of validity needed
- the approximate time and position in flight at which the request will be made
- the ATS unit the aircraft is expecting to contact.

Aircraft commanders are expected to make the maximum use of:

- VOLMET
- Automatic Terminal Information Service (ATIS) broadcasts which give all the necessary take-off and landing information for selected aerodromes.

Routine aircraft observations

These are not required in the UK FIR/UIR and the Shanwick FIR/UIR, but in the Shanwick Oceanic Control Area (OCA) they may be required. However, special aircraft observations are required in any UK FIR/UIR/OCA whenever any of the following are encountered:

- severe turbulence or severe icing
- moderate turbulence, hail or cumulonimbus clouds during transonic or supersonic flight
- other conditions which the aircraft commander considers may affect the safety of other aircraft.

Exceptionally, a meteorological office may have requested reports or there may be an agreement between the authorities and the aircraft operator to do so.

Pressure units

The ICAO standard for pressure units is the hectopascal which is identical to the millibar. The UK has notified ICAO that the millibar will continue to be used in this country.

Chapter 13
Communications

Introduction

The information for this chapter is derived mainly from the GEN section of the UKAIP. The title 'communications' is slightly misleading as it also deals with radio navigation aids.

The telecommunication services provided in the UK are divided into the following:

- *Aeronautical Mobile Service (AMS)*
  deals with the air/ground requirements of the Air Traffic Service (ATS) including the emergency services.
- *Aeronautical Radio Navigation Service (ARNS).*
- *Aeronautical Broadcast Service (ABS)*
  this refers to the meteorological broadcasts referred to in the previous chapter.
- *Aeronautical fixed service*
  this refers to the use of the Aeronautical Fixed Telecommunication Network (AFTN) and also radio, telephone and facsimile.

The Aeronautical Mobile Service (AMS)

The following points should be noted:

- the frequencies used are between 118 and 136.975 MHz. An 8.33 kHz spacing is now being introduced, providing 2280 communication channels
- procedures to follow in the event of radio failure are in the AD and ER sections of the UKAIP and are described briefly in Chapter 7
- at civil aerodromes the following words in a callsign identify an Air Traffic Control Service (ATCS) – TOWER, APPROACH, GROUND, ZONE, RADAR, DIRECTOR, DELIVERY
- INFORMATION in a callsign indicates an Aerodrome Flight Information Service (AFIS), an Aerodrome Traffic Information Service (ATIS) or an Area Flight Information Service

- in a callsign, the word RADIO identifies an aerodrome air/ground communication service.

Use of VHF R/T channels

Geographical separation is used to try and ensure that aircraft at the limits of height and range of each service and using the same or adjacent frequencies do not interfere with each other. On en-route sections these limits correspond to that of the ATC sector concerned and those for international services are:

- aerodrome (TOWER) 25 nm up to 4000 ft
- APPROACH 25 nm up to 10 000 ft.

Except in emergencies, or unless otherwise instructed by ATC, pilots should observe these limits. Services other than international services are provided on frequencies that are shared between numerous ground stations. Pilots can assist in reducing congestion by keeping communications to a minimum by limiting their use to the minimum heights and distances from the aerodromes that are operationally necessary. In the case of TOWER, Aerodrome Flight Information Service (AFIS) and air/ground facilities, restrict use, as far as possible, to 1000 ft in the immediate vicinity of the aerodrome and, in any event, to 10 nm and 3000 ft.

Common frequency for helicopter departures

A frequency of 122.95 MHz known as the DEPCOM frequency is reserved for the use of helicopters leaving a site that has no radio facilities. It is restricted to 500 ft above ground level (agl). Helicopters approaching the site should monitor the frequency but blind transmissions are not allowed.

Emergency service

An emergency service in the UK is continuously available on the international distress frequency of 121.5 MHz. It is not necessary to address the call to any particular station but prefix the call with either 'MAYDAY' or 'PAN PAN' (see the next chapter). In the event of all other means of communication having failed, dedicated satellite voice telephone numbers are available for all the UK regions.

Emergency service for aircraft on the ground

At certain aerodromes, aircraft on the ground may communicate directly with the fire service in attendance, providing contact with the ATC on the appropriate frequency is maintained. This service is only available via ATC and may only be used for the duration of the emergency. The fire service does not monitor this frequency at other times.

Aeronautical Radio Navigation Services (ARNS)

- *MF Non-directional Beacons (NDB)*
  for each beacon a range based on a daytime protection ratio between wanted and unwanted signals is published in the En-route (ENR) section of the UKAIP. Within this range, during daytime, the errors should be within ±5°. At greater ranges and particularly at night the errors could be considerably more. Beacons provided for use as approach aids in conjunction with Instrument Approach Procedures (IAP) published in the aerodrome (AD) section of the UKAIP are known as locator beacons and are indicated by (L). A pink AIC deals with the use of NDBs.
- *Very High Frequency Direction Finding Stations (VDF)*
  the bearings these give are classified as follows:
  - class A ± 2°
  - class B ± 5°
  - class C ± 10°.
  The bearings will only be given when the conditions are satisfactory and usually they will not be better than class B.
- *VHF Omni-directional Radio Range (VOR)*
  the Designated Operational Coverage (DOC) will be found in the ENR section of the UKAIP and, if the beacon is associated with an aerodrome, also in the AD section. The VORs have an automatic change-over facility in the event of failure. During the change-over period no identification signal will be transmitted as during this time inaccurate bearings may be produced.
- *Distance Measuring Equipment (DME)*
  these are usually co-located with VORs and the frequencies 'paired' so that tuning the VOR frequency automatically sets the correct UHF frequency for the DME. A simultaneous range and bearing fix can then be obtained. If the two beacons are more than 600 m apart the last letter of the callsign will be a 'Z'. Aerodrome DMEs referred to in the AD section of the UKAIP are for use in the IAPs and are usually adjusted so that zero range will be indicated when the aircraft is at the specific runway threshold.

- *Tactical Air Navigation Beacons (TACAN)*
  these are military beacons given bearings and distances to military aircraft using Ultra High Frequency (UHF) equipment. Civil aircraft can make use of the distance element by setting the aircraft VOR to the correct VHF frequency that is usually shown in brackets on the en-route charts.
- *Instrument Landing System (ILS)*
  this is a standard precision aid which gives aid during the final approach and landing. There are three categories:
  - category I, which gives guidance down to 200 ft
  - category II, which gives guidance down to 50 ft
  - category III, which using ancillary aircraft equipment, can give guidance down to 0 ft.
  Aircraft overflying the localizer, the beacon producing the approach track, or taxying on or near the ILS runway may disturb the guidance signals. ATC will take all necessary steps to prevent these problems occurring during category II or III operations. The usable coverage sector:
  - ±35° either side of the course (localiser) line
  - up to 10 nm for the glide path unless otherwise stated. This is the ICAO standard.

Reference should be made to the pink AIC which gives much useful information.

Microwave Landing System (MLS)

This is a development of the ILS working on a much higher frequency. Its main advantages are that approaches can be made from anywhere within the fan-shaped coverage area and at quite a range of glide path slopes. The azimuth coverage is ±40° up to a distance of 20 nm. The glide path approach angles can be between 0.9° and 20°.

Surveillance Radar Approach (SRA)

This is operated by the ATC using their surveillance radar. Ranges and position relative to the correct path (left/right of the centre line) will be passed to the pilot. In place of glide path information, check heights are passed. The approach may terminate at 0.5 or 2 nm, depending on the equipment being used.

Interference

Interference from portable telephones

A pink AIC deals with the problems that these can cause. Use of them in an aircraft contravenes both the aircraft and telephone users' licence conditions. Even when the telephone is in the stand-by position, they can still cause interference and should be switched off.

Interference from high-powered transmitters

Flying in the vicinity of high-powered broadcast transmitters may cause problems with aircraft electrical and electronic equipment. If a pilot experiences this interference, a ground fault report should be filed, giving full details of the occurrence. Other sources of high intensity radio transmissions are listed in the En-Route (ENR) section of the UKAIP giving the areas within which problems may occur. These areas are on the chart of airspace restrictions in the UKAIP and are also shown on en-route charts.

Potential interference from high-powered FM broadcast stations at 88–108 MHz can be expected. Gradually, restrictions will be placed on operators using ILS/VOR equipment which does not comply with the FM immunity requirements laid down in ICAO Annex10 Volume 5 (Aeronautical Radio Frequency spectrum Utilisation) and repeated in UK Airworthiness Notices. After 1 January 2001 all this equipment will have to comply with these requirements.

Emergency frequencies used world-wide

- 500 kHz   international distress frequency
- 2182 kHz   international distress frequency
- 121.5 MHz   aeronautical emergency and survival craft frequency
- 243 MHz   survival craft frequency.

Chapter 14
Search and Rescue

Introduction

Two annexes of the ICAO convention deal with these procedures:

- *Annex 11 Air Traffic Services (ATS)*
  among other things, this deals with the alerting service.
- *Annex 12 Search and Rescue (SAR).*

Contracting states all have similar arrangements so enabling ready international co-operation when the need arises. The UK has not filed any differences with ICAO affecting these procedures and so the situation in the UK will be typical of that in many other countries.

Search and rescue organisation in the UK

Figure 14.1 shows how search and rescue is organised in the UK. It can be seen that the whole area, for which the UK has taken responsibility, is divided into two Search and Rescue Regions (SRR) and each one has a Rescue Co-ordination Centre (RCC). Once there is any likelihood of an emergency the Area Control Centre (ACC) will alert the RCC who will then take over the responsibility for subsequent actions while keeping ACC informed.

Resources available to RCCs

Each RCC has the following dedicated resources at its disposal:

- Royal Air Force (RAF) and Royal Navy (RN) fixed-wing aircraft and helicopters carrying survival gear
- Department of Transport, Environment and the Regions (DETR) helicopters
- RAF Mountain Rescue Teams (MRT).

In addition, the RCC may call on the following for assistance:

- Her Majesty's Coastguard
- lifeboats
- non-dedicated RN ships and helicopters and RAF aircraft
- civil police, fire and ambulance services
- merchant vessels
- service personnel
- civilian mountain rescue teams
- British Telecom International Coast Radio Stations
- COSPAS – the Russian Cosmos Rescue System
- SARSAT – Search and Rescue Satellite Aided Tracking System
- neighbouring RCCs.

Distress frequencies carried by SAR and other military aircraft

These are given in Table 14.1.

Scene of search frequencies

SAR aircraft may use a variety of frequencies at the scene of search. The RCC will dictate which ones will be used.

Table 14.1 Frequencies used by SAR and other military aircraft

Frequency	Speech facility	Homing facility
121.5 MHz	RAF Long Range Patrol Aircraft (LRMP) and helicopters	RAF LRMP aircraft and helicopters
	Certain other military aircraft	Certain other military aircraft
	DETR helicopters	DETR helicopters
	RN Sea King helicopters	RN Sea King helicopters
243.0 MHz	RAF aircraft	RAF LRMP aircraft and
	RN aircraft	helicopters
	DETR helicopters	RN Sea King helicopters
		DETR helicopters
		Some other military aircraft
500 kHz	–	RAF LRMP aircraft
2182 kHz	RAF LRMP aircraft	RAF Sea King helicopters
	Some helicopters	DETR helicopters

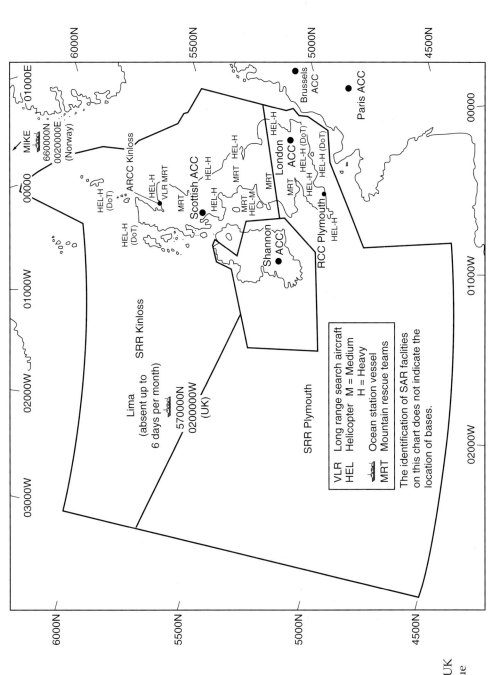

Figure 14.1 The UK Search and Rescue Organisation.

SAR agreements

As a contracting state of ICAO, the UK is committed to providing SAR services for international civil aviation throughout the defined areas on a 24-hour basis. The ICAO Regional Air Navigation Plans do not define the scale of effort which should be available but identify the required facilities which should be provided with due regard to the density of the traffic and the size and passenger capacity of aircraft operating in the region. As a member of the North Atlantic Treaty Organisation (NATO), a contracting state under the ICAO Convention and according to bilateral agreements, the UK can seek SAR assistance from the resources of other nations.

Procedures and signals used

When an ACC has reason to believe that an aircraft is in a state of emergency, it will alert the RCC and notify the local police, if appropriate. The RCC will, in turn alert SAR units and RAF Mountain Rescue Teams (MRT) and the police will notify civilian MRT, fire, ambulance and hospital services. When the location of a civil aircraft which has crashed on land is known, and no air search is necessary, the police will normally take responsibility for dealing with it. Near an aerodrome it is not easy to say where the responsibility of the SAR services begin and that of the aerodrome emergency services ends so the two must liaise closely.

Alerting

The alerting service is available for all aircraft that are known to be operating within the UK information regions. The ATSU, which was last in contact with the aircraft needing assistance, will normally start the alerting action. If a pilot intercepts a distress message then the ATSU should be informed at once. The UK RCCs do not maintain a listening watch on VHF.

Difficult areas for SAR

The UK has not formally named such areas. However, it is strongly recommended that general aviation (GA) aircraft, when proposing to operate over mountainous or sparsely populated areas, should carry appropriate survival equipment including an Emergency Locator Transmitter (ELT). If the pilot of an aircraft not equipped with radio is going to fly in one of these areas or more than 10 nm from the coast, it is essential that a flight plan is filed before departure. Flight plans can only be delivered to destinations which have access to the Aeronautical Fixed Telecommunications Network (AFTN). If this is not the case, the pilot should ensure that a responsible person at the destination is notified so that overdue action can be taken, if necessary.

General aviation flights over the North Atlantic

GA pilots flying over the North Atlantic should realise that many other countries surrounding this ocean have much more stringent requirements regarding:

- ELTs
- survival equipment
- communication equipment
- fuel reserves
- navigation equipment
- instrument ratings
- aircraft inspections.

Degrees of emergency

When an incident occurs, various organisations are alerted, the nature and state of alertness being determined by the degree of emergency. Internationally, the three phases are as follows:

- *INCERFA (uncertainty)*
  this is raised in one of the following cases:
    - an expected contact with the aircraft is 30 min overdue
    - failure to establish communication with the aircraft
    - failure to arrive within 30 min of the last established ETA.
- *ALERFA (alert)*
  this is raised when:
    - subsequent attempts fail to establish contact
    - there is no news of the aircraft
    - information is received that the aircraft's operating efficiency is impaired
    - an aircraft, cleared to land, fails to land within 5 min and there is no obvious reason for this.

- *DETRESFA (distress)*
  this is raised when:
  - further attempts to establish contact are unavailing
  - fuel is considered to be exhausted
  - information is received that the aircraft might be forced to land.

Procedure for pilot requiring SAR escort facilities

If a pilot, while flying over water or sparsely inhabited areas, believes the aircraft's operating efficiency has become impaired he or she should notify the ACC to alert the SAR organisation. It should be realised that SAR aircraft are limited in numbers and so should only be requested if absolutely necessary. When provided, the SAR aircraft will be positioned as close as possible and it is important to establish radio contact between the aircraft as soon as possible.

Flights in areas where SAR operations are in progress

Pilots should avoid these areas if at all possible. If this is not the case, pilots should:

- contact the RCC by telephone before departure
- file a flight plan showing times in the SAR area and the heights to be flown
- obtain the latest information about the weather conditions en-route and in the search area
- note that a special VHF frequency of 130.25 MHz is available for use at major emergency incidents. On this a pilot can broadcast his or her intentions as the area is approached. This is not used to augment or replace the scene of search frequencies.

As referred to in Chapter 11, in certain cases, a Temporary Danger Area (TDA) may be set up at the scene of an incident. This will usually be by NOTAM. If this fails to achieve its objective the Restriction of Flying (Emergency) Regulations will be used to make it an offence to fly in the area.

Action by survivors in life rafts

Survivors should use some or all of the following methods when searching aircraft or vessels are seen or heard:

- fire distress flares or cartridges
- use some object with a bright flat surface as a heliograph
- flash torches
- fly anything resembling a flag and, if possible, make the international distress signal by flying a round object above or below it
- use fluorescent marker to leave a trail in the sea
- use distress radios, if carried. The VHF/UHF will only have a limited life. The MF radios are hand-cranked.

Action by survivors in an isolated area

Survivors should use some or all of the following methods when searching aircraft are seen or heard:

- visual (see Figure 14.2). The symbols should be 2.5 m long and conspicuous. Make using any means – fabric, stones, trampling, staining with oil, etc.
- make the aircraft as conspicuous as possible by spreading any available material over it
- use smoke or fire
- use dinghy radio as described above.

Action by search aircraft at night

The following technique is used by RAF search aircraft at night:

- a single green pyrotechnic will be fired every 5–10 min
- after 15 seconds the survivors should fire a red pyrotechnic, followed after a short interval by a second

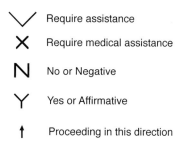

Figure 14.2 Survivors' ground signals.

• the survivors should use additional pyrotechnics if the aircraft seems to be missing them. Pyrotechnics should not be fired directly at the aircraft.

Emergency Locator Transmitters (ELT)

Those transmitting on 121.5 and 243 MHz have a distinctive warbling note repeated two to four times a second. The 121.5 MHz frequency is not exclusive to aviation; many Emergency Position Indicating Radio Beacons (EPIRB) in use on land and sea use it. Pilots listening on 121.5 MHz and hearing a warbling note should report the incident immediately to the ATCC, giving all relevant information.

The COSPAS/SARSAT satellite-aided locator beacons operated by the Russia and USA, respectively, will detect distress signals from any locator beacon and also give the location of the beacon to 20 km or better. This information will be passed to an appropriate RCC in under 90 min.

Use of old-style dinghy radios

These operate on 500 kHz when a handle is cranked. They should be used for 4 min at 10-min intervals. If accurate time is available transmissions should be made at the international silence periods of 3 min starting at 15 min past and 45 min past the hour.

Use of 121.5 MHz

Brief reference was made to this in the previous chapter and a pink AIC gives very full details which are summarised here. In the UK there is a 24-hour watch kept on this frequency and there is an ability to provide an instant fixing service. In addition, Military Emergency Diversion Aerodromes (MEDA) keep a listening watch on this frequency. If an aircraft is threatened by serious and imminent danger and is in need of immediate assistance, it is in a state of *distress* and messages should be prefixed with:

MAYDAY, MAYDAY, MAYDAY

If the pilot has a very urgent message concerning the safety of any aircraft or of a person on board or in sight, the prefix to use is:

PAN PAN, PAN PAN, PAN PAN

The message should contain as much of the following information as possible, after the above prefixes:

- name of station being addressed (if known)
- callsign and type of aircraft
- nature of emergency and pilot's intention
- present or last known position, flight level/altitude and heading
- pilot qualifications, although this is not an ICAO requirement but, in the UK, this will help the controller to plan action best suited to the pilot's ability. The prefix 'TYRO' to the controller indicates that the pilot is inexperienced.

If the message to the controller is weak or distorted, the pilot may be asked to use the speechless code. In this the pilot just presses the transmit button according to the code in Table 14.2. An alternative to this is for the pilot to operate the SSR transponder by setting codes suggested by the emergency controller.

If no state of emergency or distress exists, but the pilot is in difficulty (e.g. lost), the emergency controller may give assistance.

Practice emergencies

Pilots may simulate emergency incidents (but not the state of distress) on 121.5 MHz to gain experience of the service. Having listened out to make sure the channel is clear, the following brief message should be sent:

PRACTICE PAN, PRACTICE PAN, PRACTICE PAN, LONDON CENTRE, THIS IS GCLOT.

Unless requested by the emergency controller, the SSR Mode A 7700 (distress) should not be selected, but mode C should switched on if available.

Table 14.2 Speechless code

Number of transmissions	Meaning
One short (dit)	Yes
Two short (dit dit)	No
Three short (dit dit dit)	Say again
Four short (dit dit dit dit)	Request homing
One long (dah 2 seconds)	Manoeuvre complete
One long, two short, one long (dah dit dit dah)	My aircraft has developed another emergency

Chapter 15
Entry, Transit and Departure of Aircraft

Introduction

The rules regarding aircraft movements internationally are laid down in Annex 9 of the ICAO Convention (Facilitation). The UK notify in the UKAIP quite a number of differences from the ICAO Standards and Recommended Practices (SARPs). In general, this chapter will give the UK rules, but will note if there are any significant differences from the SARPs.

Crossing UK boundaries

In general aircraft may cross these at any point, subject to the restrictions imposed by having to follow ATS routes and avoiding prohibited, restricted and danger areas.

Restrictions on using UK aerodromes

Operators should comply with local flying restrictions and noise abatement procedures in respect of jet aircraft at UK airports, details of which are given in the AD section of the UKAIP. Care must be taken to ensure that advance arrangements have been made for the ground handling of the aircraft and that, unless special arrangements have been made, arrivals are scheduled during the normal opening hours.

Aerodrome Operating Minima (AOM)

The Air Navigation Order (ANO) does not permit aircraft registered in a country other than the UK to descend lower than 1000 feet above the aerodrome if the Runway Visual Range (RVR) is less than the minimum specified for landing on the particular runway. These minima are published in the UKAIP AD section and also on the standard Instrument Approach Charts (IAC). In addition. the aircraft shall not:

* continue an approach to landing below the specified Decision Height (DH) in the case of precision approaches
* descend below the specified Minimum Descent Height (MDH) in the case of non-precision approaches.

In either case from such a height, the specified visual reference for landing must be established and maintained. The operator is required to send details of the specific minima he will be using for category I Instrument Landing System (ILS) operations at least 7 days before the first proposed flight.

An application to conduct category II and/or category III operations, or low visibility take-offs in less than 150 m Runway Visual Range (RVR), must be accompanied by a copy of the relevant permission or certificate of competency issued to the operator by the state of registration of the aircraft.

Prevention of Terrorism Act

Under this act, aircraft carrying passengers for reward, except community air carriers with a licence granted by an European Union (EU) state, coming to Great Britain from the Republic of Ireland, Northern Ireland, Isle of Man or the Channel Islands or going to any of these places shall only land at a designated aerodrome for the purpose of landing or disembarking passengers. These aerodromes are listed in the GEN section of the UKAIP. This embargo also applies to aircraft not carrying passengers for reward.

Application may be made to the police for the area for permission to use a non-designated aerodrome.

It is also required that pilots of civil helicopters flying into Northern Ireland notify the Royal Ulster Constabulary of the time and point of arrival in Northern Ireland.

Customs and Excise and sanitary airports

Many airports in the UK have been designated as customs airports and these are listed in the AD section of the UKAIP. In addition, the major airports have been designated as sanitary airports under International Sanitary Regulations. Customs and Excise attendance at these airports is provided according to the regular traffic needs. If attendance is required outside the published hours, 'prior notice' must be given. Certain charges may be incurred in these cases.

Using non-Customs and Excise aerodromes

The restrictions for outward and return flights to many of these airports have been relaxed. The relaxations mostly apply to flights to and from EU countries and to a few non-EU flights. Prior notice of all non-EU flights and of EU arrivals (not departures) must be given to the Customs and Excise by the aerodrome manager.

Effects of the Single European Market

There is no longer any customs or immigration procedures for aircraft arriving from or departing to another EU member state. There is no restriction on the aerodromes to be used by general aviation aircraft flying to or from the EU.

Arrival in the UK from outside the EU

Without special permission, such aircraft must land at a designated Customs and Excise airport for the first time after its arrival in the UK and at any time when carrying passengers or goods brought from outside the EU and not yet cleared. On landing the aircraft must be taken to the examination station and the customs contacted for advice on the procedure to be followed. If the aircraft has stores on board, they must be declared. There is a possibility of liability to Customs and Value Added Tax (VAT):

- all civil aircraft are eligible for relief from customs duty on importation into the EU under the end-use arrangements. Details can be obtained from Customs and Excise
- aircraft under a Maximum Structural Take-off Mass of less than 8000 kg are liable to VAT at the standard rate. However, they may be imported free of VAT if:
  - the importer is a non-EU resident
  - it is intended to re-export the aircraft.

Departure to a destination outside the EU

Without special permission, all these flights must depart from a designated Customs and Excise airport. After the clearance from the designated airport, any further landing in the UK must be at a designated airport. If, however, a forced landing is necessary at a non-designated airport, the aircraft commander must immediately report to a Customs and Excise officer or a police officer. These provisions also apply to aircraft making a stopover in another EU country on its way to a destination outside the EU.

Overflying and technical stops

Provided that an aircraft is registered in a country which is party to the International Air Transport Agreement (see Chapter 1), overflying flights (First Freedom) and Technical Stops (Second Freedom) may be carried out.

Traffic stops

Foreign aircraft may not operate for hire or reward at UK airports without prior permission. This will usually be the result of a bilateral agreement between the UK and the foreign country concerned. Flights by foreign registered aircraft to take traffic between two places in the UK (cabotage) will not be permitted. Fifth freedom flights, where a foreign aircraft takes traffic from the UK and flies it to another country that is not its country of registration, will be subject to special permission.

Multilateral agreement on commercial rights of non-scheduled air services

The UK is a party to this agreement between 17, mostly western European states, which permits the exercise of traffic rights for the carriage of freight and:

- flights for humanitarian or emergency needs
- taxi services with aircraft with not more than ten passenger seats and with no resale of seats to the public
- flights on which the entire space is hired by a single person or company and is used for their staff or merchandise and no space is resold
- single flights by one operator in one month between the same two traffic centres.

Further details will be found in the CAP 555 – Overseas Non-scheduled Flight Guide.

Customs requirements for aircrew

Arriving from other EU countries the duty/tax-free allowances do not apply to intra-EU crew but no customs declaration need be made. There is no customs restriction on crew members' exit route from the airport although security measures may dictate otherwise. Arriving from outside the EU, they will either have to complete a declaration form or pass through the red customs channel if they are carrying goods in excess of their allowances. This applies also to crew arriving from outside the EU who have made a stopover inside the EU.

Departing to non-EU countries, it is not normally necessary for aircrew's effects to be made available to customs unless a refund of VAT is being claimed.

Customs requirements for passengers

There are various possibilities:

- *Arriving on domestic flights*
  the hold baggage of passengers who originally departed from a non-EU country and have then transferred to a domestic flight will be subject to customs control if this has not already been done at the first airport of arrival.
- *Arriving from other EU countries*
  no customs declaration is necessary and there is often a separate exit for them. If however, they started their journey from outside the EU and have transferred to the UK flight in another EU country, they must go through the full customs procedure.
- *Arriving from non-EU countries*
  passengers must go through the full customs procedure unless they are transferring to another flight to a non-EU country.
- *Departing to any country*
  there is no need to submit baggage to customs unless a refund of VAT is being claimed.

Aircrew immigration requirements

A member of aircrew arriving or departing should have a valid crew licence or certificate that includes a certification that the holder may at all times re-enter the issuing state. This is acceptable as an Identification Document (ID). The crew member may arrive on one aircraft and depart on that one or another within 7 days unless:

- there is in force a deportation order against the crew member
- the crew member has been previously refused leave to enter and this has not been rescinded
- an Immigration Officer requires the crew member to submit to examination.

Passenger immigration requirements

For immigration purposes the UK, the Channel Isles, the Isle of Man and Eire together form a Common Travel Area (CTA). A person arriving from or departing to places outside the CTA is liable to be examined by an Immigration Officer and may be required to produce a valid passport or some other acceptable document establishing identity and nationality or citizenship, endorsed when necessary with a current UK visa or entry clearance.

Obligations of captains, agents or owners of aircraft under Immigration Act

They must ensure that:

- passengers are only processed at a designated port of entry
- passengers do not disembark except with the approval of the Immigration Officer
- passengers are presented in an orderly manner for examination
- if required a list of crew and passengers is provided
- all passengers who are not citizens of the European Economic Area (EEA) are provided with landing cards; the EEA refers to the EU and Iceland and Norway
- passengers that are refused entry are removed from the UK
- passengers placed on the aircraft under the authority of an Immigration Officer are detained in custody
- they pay the Secretary of State any expenses incurred as the result of an illegal immigrant having to be detained
- a person against whom a deportation order is in force is removed from the UK.

Immigration (Carriers' Liability) Act

Under this a charge of £2000 will be levied on carriers who bring to the UK any single passenger that has no proper documentation on arriving in the UK .

Public health requirements

The airport medical officer may examine the following persons and take any necessary measures for preventing danger to the public health:

* if the person is suspected of suffering from, or to have been exposed to infection from an infectious disease or suspected of being verminous
* if the person is departing from the UK and there are reasonable grounds for believing the person is suffering from a quarantinable disease subject to international health regulations.

Experienced holders of non-UK professional licences

If these persons wish to fly a UK registered aircraft in a private capacity and exercise the privileges of an Instrument Rating (IR), the CAA will be prepared to consider granting a certificate of validation permitting this to be done.

Visiting pilots – instrument and flying instructor ratings

Pilots holding valid licences including an IR and flying instructors rating issued by an ICAO contracting state will need to obtain at least a UK Private Pilot's Licence (PPL) if they wish to fly under Instrument Flight Rules (IFR) in controlled airspace (CAS) or to give flying instruction.

General declaration

This is a document showing details of the flight and a declaration of health including details of any disinsecting or sanitary treatment that has been carried out. These documents are not now required for flights within the EU but are usually required for flights to the rest of the world. The number of copies required varies considerably. It is never less than six but some countries require many more.

Importation into the UK of mammals

With the exception of some farm animals and horses, this is strictly forbidden. All mammals being imported have to be declared and they will then have to be quarantined at the expense of the importer. Proof of vaccination has no relevance under the present regulations. A white AIC gives more details.

Chapter 16
Aircraft Registration and Airworthiness (ICAO Annex 8)

Introduction

Most of this information is taken from the Air Navigation Order (ANO) and the numbers quoted after ANO refer to the 1995 order. If a later edition is published there will be a table of comparisons showing the cross-references to the 1995 order. The UK follows ICAO Standards and Recommended Practices (SARPs) and references will be made to the relevant ICAO Annex.

Joint Aviation Authority (JAA) publications

These cover much of the same ground as the ANO. Eventually all the members of the JAA will subscribe to their rulings. In this chapter, reference will be made to both the ANO and JAA publication where relevant. The JAA regulations (JAR) referred to will be:

- *JAR–OPS Parts 1 and 3*
  this refers to commercial air operations for aeroplanes and helicopters, respectively.
- *JAR-FCL Parts 1 and 2*
  this refers to the flight crew licensing for aeroplanes and helicopters, respectively.
- *JAR-FCL Part 3*
  this refers to medical requirements for all pilots.
- *JAR 145 Approved Maintenance Organisations*
  gives the requirements for granting maintenance approval.
- *JAR-TSO Joint Technical Standard Orders.*
- *JAR-23 Normal, Utility, Aerobatic and Commuter Category Aeroplanes*
  airworthiness standards for the first three of these categories having passenger seating for not more than nine and a Maximum Structural

Take-off Mass (MSTOM) of 12 500 lb (5700 kg) or less. Also propeller-driven twin-engined aeroplanes in the commuter class having passenger seating for not more than 19 passengers and a MSTOM of up to 19 000 lb (8600 kg).

- *JAR-25 Large Aeroplanes*
  airworthiness standards for multi-engined turbine powered aeroplanes with a MSTOM over 5700 kg.
- *JAR-27 Small Rotorcraft*
  airworthiness standards for rotorcraft of MSTOM of up to 6000 lb (2700 kg).
- *JAR-29 Large Rotorcraft*
  airworthiness standards for rotorcraft of MSTOM over 6000 lb (2700 kg).

Aircraft to be registered – ICAO Annex 7 and ANO 3

With certain exceptions in respect of gliders and aircraft being flown for test or experimental flying ('B' conditions), aircraft shall not fly in or over the UK unless registered in:

- some part of the Commonwealth
- an ICAO contracting state
- a country having an agreement with the UK.

Registration of aircraft in the UK – ANO 4 and Schedules 1, 2 and 3

- *Paragraph 6 of Article 4*
  a written application must be made to the CAA for an aircraft to be registered in the UK including proof of ownership and any chartering of the aircraft. It is important that a proper description of the aircraft is given according to the 'General Classification of Aircraft' set out in Schedule 1 Part A of the ANO.
- *Paragraph 10 of Article 4*
  if after registration, an unqualified person becomes owner or part-owner of an aircraft, the registration is void and the Certificate of Registration must be returned to the CAA.
- *Paragraph 11 of Article 4*
  a registered owner must tell the CAA if:
  - there is any change in the information originally supplied
  - the aircraft is destroyed or permanently withdrawn from use
  - the demise charter is ended.

Air Operator's Certificate (AOC) – ANO 6 and JAR-OPS 1.175

An aircraft shall not fly for the purpose of public transport unless the CAA has issued an AOC. Before issue, the CAA will need to be satisfied with the operator, the operator's organisation, staff, maintenance and the arrangements to secure the safe operation of the approved aircraft in the approved areas. An essential part of the procedure will be the preparation and approval of an Operations Manual (OM) with which all operations will have to comply.

Certificate of Airworthiness (C of A) – ICAO Annex 8, ANO 8 and 9

An aircraft is prohibited from flying without a valid C of A which will include the Aeroplane Flight Manual (AFM), unless it is flying wholly within the UK, and it is:

- a glider not being used for the public transport of passengers or aerial work apart from flying instruction or testing
- a balloon on a private flight
- a kite
- an aircraft flying under 'A' conditions (mainly for the renewal of the C of A) or 'B' conditions (mainly for experimental and test flying by manufacturers) an aircraft issued with a permit to fly under certain conditions.

The C of A is issued by the authority when it is satisfied that it fulfils all the requirements laid down for the particular type of aircraft and it has been maintained satisfactorily. Airworthiness requirements are laid down in British Civil Airworthiness Requirements (BCARs) for older aircraft and under the appropriate Joint Airworthiness Requirements (JAR) (see the earlier section 'JAA publications').

C of A Aircraft categories – ANO Schedule 3

C of As will specify the category of the aircraft and this will determine the purposes for which it may be flown. These are:

- *Transport category (passenger)*
  any purpose.
- *Transport category (cargo)*
  any purpose except public transport of passengers.

- *Aerial work category*
  any purpose other than public transport.
- *Private category*
  any purpose other than aerial work or public transport.
- *Special category*
  any purpose other than public transport specified in the C of A but not including the carriage of passengers unless expressly permitted.

Certificate of Maintenance Review – ANO 10

A UK-registered aircraft, having a C of A, shall not fly unless the aircraft and all its equipment is maintained in accordance with a maintenance schedule approved by the authority in relation to that aircraft. If the C of A is issued in the transport or aerial work category, the aircraft will also require a Certificate of Maintenance Review. The following apply to this certificate:

- the maintenance schedule will specify when these reviews must be carried out
- the certificates can only be issued by a licensed aircraft engineer or other person authorised by the authority to do so
- the certificate will only be issued when all the appropriate maintenance and inspection procedures have been carried out, any defects in the technical logs have been rectified and all necessary Certificates of Release to Service have been issued
- two copies of the Certificate of Maintenance Review are prepared; one is retained by the operator and one carried in the aircraft
- the certificate should be retained for 2 years.

Technical log – ANO 11 and JAR-OPS 1 1.915

This must be carried in any UK aircraft registered in the transport or aerial work category. At the end of every flight, the aircraft commander shall enter in the technical log:

- the times of take-off and landing
- particulars of any defects – zero report is also required
- any other particulars regarding airworthiness or operation of the aircraft that the authority may require.

If the Maximum Structural Take-off Mass (MSTOM) does not exceed 2730 kg and it is not operated by the holder of an AOC, the entries may be made in any other authority-approved record.

In the case of several consecutive flights on the same day and under the same commander, the entries may be made at the end of the last flight if a defect did not occur during an earlier flight. If these consecutive flights are for public transport, the flights must start and end at the same aerodrome. Technical logs must be preserved until a date 2 years after the aircraft has been destroyed or permanently withdrawn from service.

Certificate of Release to Service – ANO 12

This is a certificate showing that part of the aircraft or its equipment has been overhauled, repaired, replaced or modified in an approved manner and with approved materials. Persons entitled to issue these certificates:

- holders of an aircraft engineer's licence
- persons approved by the authority
- in the case of adjusting and compensating direct-reading compasses holders of Airline Transport Pilot's Licence (Aeroplanes)
- persons approved under JAR-145 and in accordance with that approval.

If the procedure cannot be done in a manner qualifying for the issue of a certificate or at a place where a certificate cannot be granted, the aircraft may fly to a place where this can be done. The pilot must take into account flight safety, equipment availability, liberty and health of anyone on board and then report the details of the flight to the authority within 10 days.

Equipment of Aircraft – ANO 14 and 15, Schedules 4 and 5, JAR-OPS 1.875

The requirements for communication and navigation equipment, as set out in JAR-OPS 1, is as follows:

- approved and installed as required and must have the minimum performance standards and satisfy the operational and airworthiness requirements
- must be installed so that the failure of any single unit required for navigation or communication purposes does not result in the failure of another unit required for these purposes

- in operable condition for the kind of operation being planned except as provided in a Minimum Equipment List (MEL); see the next section
- all equipment must be readily operable by any crew member needing to use it
- all equipment must have the minimum performance standards, as laid down in JAR-TSO.

Schedules 4 and 5 of the ANO give the equipment required in considerable detail for all types of aircraft and operations.

Minimum Equipment List (MEL) – ANO 16 and JAR-OPS 1.030

With the authority's permission, an aircraft may start a flight even if all the required equipment is not being carried or is unserviceable, providing the aircraft commander is satisfied that the flight can be conducted in safety. An aircraft will have an approved MEL giving the items that can be unserviceable, but a flight can still be carried out. The MEL quite frequently will impose limitations on the operations. The purpose of the flights is for the aircraft to get to a base where the necessary repairs or replacements can be made. It is not intended to be a long-term permit to fly with a deficiency. A copy of the MEL is often carried in the Operations Manual (OM) and must be carried if it is being made use of.

Aircraft, Engine and Propeller Log Books – ANO 17 and Schedule 6

For UK registered aircraft the following must be kept:

- an aircraft (airframe) log book
- a log book for each engine
- a log book for each variable pitch propeller.

If aircraft have a Maximum Structural Take-off Mass (MSTOM) not over 2730 kg, the log books must be of a type approved by the authority. Other aircraft log books must include the particulars laid down in Schedule 6 of the ANO.

Each entry must be made as soon as possible and in any event not more than 7 days after the Certificate of Maintenance Review expires. If other documents are referred to in the log book they must be clearly identified and will be regarded as part of the log book. Entries have to be made every time any maintenance, overhaul, repair, replacement, modification or inspection is carried out.

All log books need to be kept until 2 years after the aircraft, engine or propeller has been destroyed or withdrawn from use.

Aircraft Mass Schedule – ANO 18 and JAR-OPS 1.605 Appendix

Every flying machine or glider must be weighed and its centre of gravity (CG) found, as required by the CAA. It is quite usual for this to be done when a new C of A is about to be issued. After the aircraft has been weighed a mass schedule has to be prepared showing:

- the basic mass of the aircraft, that is the empty aircraft including unusable fuel and oil and such other items as the operator wishes to have included or such other mass as approved by the authority
- the CG for the mass obtained.

The mass schedule should be retained for 6 months after the next occasion on which the aircraft is weighed.

Chapter 17
Flight Personnel

Introduction (ICAO Annex 1 – Personnel Licensing)

As in the previous chapter, reference will be made to the ANO and to the JAR-OPS and reference will also be made to:

- JAR-FCL Part 1 Flight Crew Licensing (Aeroplanes)
- JAR-FCL Part 2 Flight Crew Licensing (Helicopters)
- JAR-FCL Part 3 Flight Crew Licensing (Medical).

Flight Crew Composition – ICAO Annex 1, ANO 20, JAR-OPS 1.940

JAR-OPS 1 requires that:

- the composition of the flight crew and their numbers is no less than laid down in the Aeroplane Flight Manual (AFM), which is an integral part of the C of A
- additional flight crew as required for the particular operation and total numbers are not less than specified in the Operations Manual (OM)
- all flight crew hold applicable and valid licences and are suitably qualified and competent to conduct the duties assigned to them
- procedures are set up to prevent the crewing together of inexperienced flight crew members
- one pilot, suitably qualified in accordance with JAR-FCL, is named as the commander. The commander is permitted to delegate the conduct of the flight to another suitably qualified pilot
- when a dedicated system panel operator is required by the AFM, the flight crew includes a crew member suitably qualified as a flight engineer
- a flight crew member that is self-employed and/or working on a freelance or part-time basis complies with all the regulations regarding working on a number of different aircraft types or variants.

If operations are to be carried out under IFR or at night:

- on a turbo-prop aeroplane with a maximum passenger seating capacity of more than nine and for all turbojets, the minimum flight crew is two pilots
- other aeroplanes may be operated by one pilot provided that:
  - a recurrent training programme dealing with single pilot operations is in the OM and is carried out on the type of aeroplane to be flown
  - the pilot has a minimum of 50 hours on the aeroplane type under IFR and 10 hours as commander
  - recent experience shall be five IFR flights, including three instrument approaches, during the past 90 days on the class of aeroplane concerned.

Conversion Training and Checking – JAR-OPS 1.945

An operator shall ensure that:

- a flight crew member completes a type rating course when changing from one type of aeroplane to another
- a flight crew member completes a conversion course before commencing unsupervised line flying when necessary
- conversion training is in accord with a detailed syllabus included in the OM and approved by the authority
- the amount of training is suggested by previous training records
- before commencing line flying under supervision, approved operator proficiency checks are carried out. These are required every 6 or 12 months if two tests are carried out 3 months apart
- on completion of line flying under supervision, an approved line check is carried out to check competence at carrying out normal line operations as laid down in the OM. This check has a 12-month validity or 24 months if two checks are carried out within 9 months
- once a conversion course has been started, no other flying duties on another type must be undertaken
- Crew Resource Management (CRM) must be incorporated in the conversion courses. Pink AICs deal with this in greater detail various of these tests may be combined.

Differences and Familiarisation Training – JAR-OPS 1.950

An operator shall ensure that flight crew complete:

- differences training when operating different aeroplanes or when changing equipment and/or procedures on types currently operated, using the aeroplane or an appropriate training device
- familiarisation training which requires additional knowledge when operating another aircraft of the same type and when changing equipment/or procedures on types or variants currently operated.

Nomination as Commander – JAR-OPS 1.955

An operator shall ensure that for upgrade to commander from co-pilot and for those joining as commanders:

- a minimum level of experience is laid down in the OM
- for multi-crew operations, the pilot completes a command course. This must be specified in the OM and include:
  - training in a flight simulator including Line Orientated Flying Training (LOFT)
  - commander's responsibilities
  - proficiency check on operating as a commander
  - line training under command and under supervision for a minimum of ten sectors for pilots already qualified on the type
  - completion of a commander's line check
  - CRM.

Commanders Holding a Commercial Pilot's Licence (CPL) – JAR-OPS 1.960

An operator shall ensure that a CPL holder does not operate as a commander of an aeroplane certificated in the Aeroplane Flight Manual (AFM) for single pilot operations unless:

- when conducting a flight under VFR more than 50 nm from the departure aerodrome, the pilot has a minimum of 500 hours total flight time on aeroplanes or holds a valid Instrument Rating (IR)
- when operating on a multi-engined type under IFR, the pilot has a minimum of 700 hours flight time on aeroplanes which includes 400 hours in command of which 100 hours have been under IFR, including 40 hours multi-engined operation. The requirements of JAR-OPS 1.940 (see the earlier section) are satisfied
- a command course has been completed.

Recurrent Training and Checking – JAR-OPS 1.965

An operator shall ensure that:

- each flight crew member does recurrent training and checking relevant to the aircraft being operated
- the programme is set out in the OM and approved by the authority
- suitably qualified personnel are used for all the training and checking.

In addition, an operator shall ensure that each flight crew member does an operator proficiency check, which used to be known as the base check. The following points refer:

- the object is to demonstrate competence in carrying out normal, abnormal and emergency procedures
- the check is conducted without external visual reference and under IFR
- each flight crew member will be checked as part of a normal flight crew complement
- the validity is 6 calendar months in addition to the month of issue.

Each flight crew member will also have to do:

- *A line check*
  to show competence to carry out the normal line procedures as laid down in the OM. The validity is 12 months in addition to the remainder of the month of issue. If issued within 3 months of the date of expiry of the previous line check, the period of validity will extend to 12 months from the previous expiry date.
- *Emergency and safety equipment training and checking*
  the period of validity is the same as for the line checks.
- *Ground and refresher training*
  the period of validity is the same as for the line checks.
- *Aeroplane/flight simulator training*
  the period of validity is the same as for the line checks.

Pilot Qualification to Operate in Either Pilot's Seat – JAR-OPS 1.968

An operator shall ensure that a pilot who may be required to operate in either seat completes appropriate training and checking as specified in the OM which will have been approved by the authority.

Recency – JAR-OPS 1.970

Pilots must have done the following in the past 90 days. This period may be extended to 120 days by line flying under the supervision of a type rating instructor:

- *Commander*
  three take-offs and landings in the aircraft type or an approved simulator.
- *Co-pilot*
  if the co-pilot is to handle the controls during take-off and landing, he or she must have served at the controls during take-off and landing.

Route and Aerodrome Competence – JAR-OPS 1.975

Before a pilot is to act as a commander or as a pilot to whom the conduct of the flight may be delegated by the commander, he or she should have obtained adequate knowledge of the routes to be flown and the aerodromes that may be used. The period of validity is 12 calendar months in addition to the month of qualification. If revalidated within the final 3 months, the period will be extended to 12 months from the expiry of the last previous validation.

Operation on More than One Type or Variant – JAR-OPS 1.980

An operator shall ensure that pilots do not operate on more than one type or variant unless they are competent to do so, taking account of:

- the level of technology
- operational procedures
- handling characteristics.

An operator shall set out in the OM the procedures and/or operational restrictions for any pilot operating more than one type or variant considering:

- the pilot's minimum experience level
- the minimum experience level on one type before training for another
- the process whereby pilots qualified on one type will be trained on another
- all applicable recent experience for each type or variant.

Operation of Helicopters and Aeroplanes – JAR-OPS 1.981

When a pilot is to operate both helicopters and aeroplanes, he or she must only be allowed to operate one of each type.

The procedures and/or operational restrictions involved must be set out in the OM.

In-flight Relief of Flight Crew Members – JAR-OPS 1.940 Appendix 1

A flight crew member may be relieved of duties at the controls by another suitably qualified flight crew member. A commander may delegate the conduct of the flight to another qualified commander or if above FL 200 to a pilot that satisfies the following requirements:

- valid Airline Transport Pilot Licence (ALTP)
- appropriate conversion and checking for the type
- all appropriate recurrent training and checking
- route competence qualifications as described in an earlier paragraph.

A co-pilot may be relieved by a suitably qualified pilot or a cruise relief pilot satisfying the following requirements:

- valid Commercial Pilot Licence (CPL)
- appropriate conversion training and checking including type rating, except the requirements for take-off and landing training
- all appropriate recurrent training and checking except for take-off and landing training
- only to operate in cruise and not below FL 200
- route recency experience is not required, but flight simulator recency and refresher skill training should be carried out at intervals not exceeding 90 days.

A system panel operator may be relieved in flight by a crew member who holds a flight engineer's licence or by a flight crew member with a qualification acceptable to the authority.

Flight Crew Licences – JAR-FCL

It is the intention of the CAA to accept fully all the requirements of JAR-FCL and the UK system of flight crew licensing will cease to exist. The licences issued by the CAA after 1 January 2000 will comply with the JAR criteria. Holders of these new licences will be entitled to operate in aircraft registered in any of the member states of the JAA. Table 17.1 gives a brief summary of the licences that will be issued and their privileges. It will be appreciated that there is much greater detail than is apparent in this table and reference to the numerous white AICs and to the JAR-FCL documents will be necessary to obtain this.

JAR ratings and checks

Table 17.2 gives a brief summary of all these that are required for fixed-wing pilots.

Cabin Crew – JAR-OPS 1.888

An aircraft shall not be operated if the maximum approved passenger seating exceeds 19 and at least one passenger is actually being carried, unless at least one cabin crew member is on board for the purpose of performing the duties laid down in the OM for the safety of passengers. The minimum cabin crew should be at least one for every 50, or fraction of 50, passenger seats. Under exceptional circumstances the authority may require an operator to increase this number. In unforeseen circumstances, the required number may be reduced, provided that the number of passengers has been reduced in accordance with procedures laid down in the OM and a report is sent to the authority after the flight.

If the OM lays down greater numbers, they will be the legal requirement for that operator.

Minimum Requirements for Cabin Crew – JAR-OPS 1.995 *et seq.*

An operator shall ensure that each cabin crew member:

- is at least 18 years of age
- is medically fit to carry out the duties laid down in the OM
- remains medically fit for these duties
- is competent to carry out the duties laid down in the OM and has completed an approved initial training course
- has completed conversion and differences training before being assigned to different aircraft
- after training has completed familiarisation flights as a supernumary to the minimum crew required

Table 17.1 Pilot's licences

JAR-FCL or CAP ref	Licence	Age limits	Medical class	Flying hours to qualify
1.085	Student	16+ on first solo	1 or 2	
1.100 to 1.135, CAP 53 2/1	PPL, Landplanes	over 17	1 or 2	35 hours (FCL) 40 hours (CAP)
1.140	CPL (A), Aeroplanes	18–60	1	150 on integrated course, 200 on modulated course
CAP 54, Chapter 7	CPL (H), Helicopters	over 18	1	150 (100 helicopter)
CAP 54, Chapter 10	CPL(B), Balloons, unrestricted	over 18	1 or 2	75 at least 60 as pilot in charge (PIC)
CAP 54, Chapter 10	CP(B), Balloons, restricted	over 18	1, 2 or 3	35 at least 20 as PIC
CAP 53 2/2	PPL (Microlight)	over 17	1, 2, 3 or personal declaration of health*	15 at least seven as PIC
CAP 53 2/4	PPL (SLMG)	over 17	1, 2, 3 or personal declaration of health*	40 at least 10 as PIC
CAP 54 2/5	PPL (powered parachute)	over 17	1, 2, 3 or personal declaration of health*	4 at least 1 as PIC
CAP 54 2/9	CPL (Airships)	over 18	1	Not issued†
ANO Schedule 8, A4	CPL (Gliders)	over 18	1	British Glider Association Course
1.265 to 1.295	ATPL (A) Aeroplanes	over 21–60	1	1500
1.060	CPL (A) and ATPL (A)	60–64	1	As above
CAP 54, Chapter 8	ATPL(H)	over 21	1	1200 as pilot of flying machines including 400¶ as PIC (H)

*Signed by AME or GP.
†Check with CAA.
‡Over 65 no public transport.
¶Or various other combinations.

Entitled to fly, as for public transport (provided instrument and type rated)	
Co-pilot – restrictions	Captain – restrictions

All flying authorised by instructor

No PT or Aerial Work (AW) except as flying instructor under club auspices

None	Aircraft certificated for one pilot operation
None	Aircraft certificated for one pilot operation

None

Private and AW only

no PT or (AW) except as flying instructor under club auspices

no PT or (AW) except as flying instructor under club auspices

None

None

None

Only as a member of a multi-pilot crew and then the only one over 60‡

Table 17.2 JAR requirements of pilot's ratings and checks

JAR reference	Rating/certificate	Validity	Privileges	Tests required	Notes
FCL 1.245(a) and (b)	Type Rating (TR), Multi-engine class rating	1 year	To act as pilot on the type	Skill test conducted by a TR examiner (TRE)	If rating is restricted to acting as co-pilot this will be noted
FCL 1.245(c)	Type Rating (TR), Single pilot type/single engine rating	2 years	To act as pilot on the type	Skill test conducted by a TR examiner (TRE)	
FCL 1.185	Instrument Rating (IR)	1 year	To fly aeroplanes under IFR with minimum decision height of 200 ft	Skill test conducted by a TRE	Decision heights of less than 200 ft may be authorised after further training and testing
FCL Part 3	Medical class 1 for age over 40	6 months	To hold a CPL or ATPL	Exam by Authorised Medical Examiner (AME)	White AIC gives further details
OPS 1.035	Medical class 1 for age under 40	1 year	To hold a CPL or ATPL	Exam by AME	White AIC gives further details
FCL Part 3, OPS, 1.095/1.145	Medical class 1 or 2	Certificate gives period	To hold a PPL or to fly solo as a student pilot	Exam by AME	
OPS 1.965(d)	Emergency and safety equipment training	1 year*	To fly on aircraft type	Test by operator's staff	Also applies to cabin crew
OPS 1.965(b)	Operator's proficiency check	6 months†	To fly on aircraft type	Flight or simulator test by check pilot	No passengers. Check on emergency procedures

OPS 1.965(c)	Line check	1 year*	To fly on line	On normal flight with check pilot	
OPS 1.970	Commanders's recency	90 days‡	To fly on type	To have done three take-offs and landings in last 90 days	May be done in simulator
OPS 1.970	Co-pilot's recency	90 days‡	To fly on type	To have done one take-off and landing in last 90 days	May be done in simulator
OPS 1.965(e)	Crew Resource Management (CRM)	As scheduled	To continue operating	Training by management	
OPS 1.965(g)	Aeroplane/simulator training	1 year*	To continue operating	Training by management	
OPS 1.975(a)	Route and aerodrome competence for any pilot that may act as commander	1 year*	To continue operating in the areas		Revalidation by operating on the routes or to the aerodromes
OPS 1.965 (f)	Ground and refresher training	1 year*	To continue operating	Training by management	

*Another test in 9 months will extend validity to 2 years.

†Another test in 3 months will extend validity to 1 year.

‡May be extended to 120 days by line flying under type rating examiner or instructor.

- carries out recurrent training particularly covering the emergency procedures and drills every 12 months. Every 3 years much more realistic training is required where actual emergencies are simulated. As with the flight crew, the 12-month period can be extended to 2 years if the recurrent training is carried out 3 months before the due expiry date
- if absent from flying duties for over 6 months special procedures are set out as regards refresher training
- after all the training sessions referred to above, a check should be made to verify the crew member's proficiency. These checks must be carried out by approved personnel
- cabin crew are normally restricted to not operating on more three aircraft types. Four is usually acceptable, if the safety equipment and procedures for at least two of the aircraft are similar.

Crew Member's Responsibilities – JAR-OPS 1.085

A crew member shall be responsible for the proper execution of duties that are related to the safety of the aeroplane and its occupants and are laid down in the OM. In addition a crew member shall:

- report to the commander any incident that has, or may have, endangered safety
- make use of the operator's incident-reporting schemes

A crew member shall *not* perform duties on an aeroplane:

- while under the influence of any drug that may affect his or her faculties dangerously
- until a reasonable time has elapsed since deep-sea diving
- until a reasonable time after donating blood
- if in any doubt of being able to perform the required duties or is suffering from fatigue or feeling unfit to the extent that the flight may be endangered
- if alcohol has been consumed within 8 hours of reporting for flight duty or if the blood alcohol level exceeds 0.2 promille or if alcohol has been taken during the flight duty period or while on standby.

The Commander's Responsibilities Pre-flight – JAR-OPS 1.290 and ANO 38

Before commencing a flight the commander should ensure:

- that an operational flight plan has been prepared
- the aeroplane is airworthy and is not going to be operated contrary to the Configuration Deviation List (CDL)
- the instruments and equipment legally required are on board and serviceable unless permitted otherwise by the Minimum Equipment List (MEL)
- the sections of the OM required are on board
- the documents required to be carried are on board (see Chapter 19)
- current maps and associated data required for the route are available
- ground facilities and services required for the route are available and adequate
- the OM provisions for fuel, oil, oxygen, minimum safe altitudes (MSA), Aerodrome Operating Minima (AOM) and availability of alternate aerodromes can be complied with
- the load is properly distributed and secured
- the mass at the beginning of take-off roll satisfies the appropriate performance requirements
- all other operational limits can be complied with
- that the flight can be safely made considering the weather forecast
- that the Certificate of Maintenance Review is valid and will remain so during the flight
- in the case of an airship or balloon that sufficient ballast is being carried.

The Commander's Responsibilities from Embarkation – JAR-OPS 1.085 (e)

The commander shall:

- be responsible for the safe operation of the aeroplane and its occupants during the period of the flight
- have authority to give all orders he thinks necessary to secure the safety of the aeroplane and of the persons and/or property on board
- have authority to disembark any person or any part of the cargo which he or she thinks may represent a potential danger
- not allow a person to board who appears to be under the influence of drugs or alcohol so as to be a danger to the safety of the aircraft or its occupants
- have the right to refuse transportation of any inadmissible persons, deportees or persons in custody if their carriage poses any risk to the safety of the aeroplane or its occupants
- ensure that all passengers are briefed on the location of emergency exits and the location and use of relevant safety and emergency equipment

- ensure that all operational procedures and check lists are complied with as laid down in the OM and that the pre-flight inspection has been done.
- not permit any crew member to perform any activity during take-off, initial climb, final approach and landing except those required for the safe operation of the aircraft
- not permit any flight data recorder (FDR) to be interfered with during flight and not permit data to be erased after flight in the event of an accident or incident subject to mandatory reporting
- not permit a cockpit voice recorder (CVR) to be disabled or switched off during flight unless he or she believes that the data, which would otherwise be automatically erased should be preserved for incident or accident investigation nor to be manually erased after flight in these cases
- in an emergency that requires immediate decision and action, take any action considered necessary. The rules and operational procedures may be deviated from in the interests of safety.

Admission to Flight Deck – JAR-OPS 1.100

An operator must ensure that no person other than a flight crew member, assigned to the flight, is admitted to or carried on the flight deck unless:

- an operating crew member
- an authority representative responsible for certification, licensing or inspection
- permitted by and carried in accord with instructions in the OM.

The commander shall ensure that:

- in the interest of safety, admission to the flight deck does not cause distraction and/or interfere with the flight's operation
- all persons carried on the flight deck are familiar with the relevant safety procedures
- the final decision regarding the admission to the flight deck is the commander's responsibility.

Crew Fatigue – Application and Interpretation – ANO 62

This is a very vexed subject which will eventually be the subject of JAR rules which have still to be issued. Until such time the national guidelines set out for the UK in CAP 371 'The Avoidance of Fatigue in Aircrews' will have to be followed. The following general points should be noted:

- the regulations in the ANO refer to UK-registered aircraft that are engaged in public transport operations but they do not apply to flying instruction
- 'flight time' is the time spent, while in flight, by a crew member in an aircraft (registered in the UK or not) except for aircraft under a MSTOM of less than 1600 kg which is not flying for public transport or aerial work
- 'day' is 24 hours from midnight UTC (GMT).

The expression 'in flight' means:

- for a fixed-wing aircraft, from when the aircraft moves under its own power until it comes to rest after flight
- for a helicopter, from the first movement under its own power for the purpose of take-off until the rotors are stopped.

Crew Fatigue – Operator's Responsibilities – ANO 63

The operator must ensure that the Flight Time Limitations (FTL) laid down in the OM are observed and ensure that the aircraft does not fly unless:

- a FTL scheme has been established which has been approved by the authority which is in the OM or a copy is made available to all crew members
- all practical steps have been taken to ensure crew members' compliance with the scheme
- the operator is sure that no crew member is likely to suffer from fatigue and endanger the aircraft and its occupants
- an accurate and up-to-date record of the flying times and the duties performed during these times by each crew member is maintained.

An operator must keep these records for 12 months after a flight. To obtain an Air Operator's Certificate (AOC) the operator's FTL scheme must be approved by the authority.

Crew Fatigue – Responsibility of Crew – ANO 64

A member of a crew on any aircraft engaged in public transport shall not fly unless:

- he or she believes that he or she is not, and will not, suffer from fatigue so that the aircraft and occupants will be endangered
- he or she has made the operator aware of all flight times during the 28 days preceding the flight.

Maximum Flight Times for Flight Crews – ANO 65

Flight crew members must not commence a flight if their flying times exceed:

- 100 hours in the 28 days expiring at the end of the day on which the flight begins. In other words, 100 hours in the preceding 27 days. The fact that during the flight this limit may be exceeded is not significant
- 900 hours during the 12 months to the end of the previous month. Realistically, this means that a pilot must not exceed an average of $900 \div 13 = 69$ hours per 28 days.

These limits do not apply if a flight is made:

- in an aircraft whose MSTOM is not over 1600 kg and is not flying for public transport
- in an aircraft not flying for public transport or operated by an air transport undertaking, if at the time when the flight begins, the aggregate flight times of the flight crew member since the last medical examination do not exceed 25 hours.

Action if Medically Unfit – ANO 22

All flight crew licence holders must inform the Authority in writing in case of any injury or illness as follows:

- *Personal injury*
  as soon as possible.
- *Illness*
  after a period of 20 days.

Chapter 18
Operation of Aircraft (ICAO Annex 6 – Operation of Aircraft)

Introduction

Much of this information is taken from ANO Part V together with relevant information from JAR-OPS. Much of it will be dealing with public transport operations with which JAR-OPS is particularly concerned.

Operations Manual (OM) – ANO 27 and JAR-OPS 1 Appendix to 1.1045

The JAR-OPS goes into the detail of the OM contents at great length as also does CAP 360 – Air Operators Certificates. The ANO Schedule 10 Part A gives a brief indication of its contents. The subjects listed in the JAR-OPS are as follows:

- crew composition
- crew qualification requirements
- crew health precautions
- Flight Time Limitations (FTL)
- operating procedures
- dangerous goods and weapons
- security
- handling of accidents and occurrences
- rules of the air
- aeroplane limitations
- normal, abnormal and emergency procedures
- performance data
- flight planning data and procedures
- mass and balance and loading
- Configuration Deviation List (CDL) and Minimum Equipment List (MEL) and their use
- survival and emergency equipment including oxygen

- emergency evacuation procedures
- aeroplane systems including operating instructions
- route and aerodrome instructions and information – an outside agency will usually supply these
- training – details of syllabi and checking procedures.

In the UK the requirement for the OM applies to all public transport aircraft except when on a flight of less than 1 hour for the purpose of training or for any purpose providing the flight begins and ends at the same aerodrome. The operator must have procedures for keeping the OM up-to-date and for ensuring that everyone concerned with operations both flight and ground staff have ready access to all sections of the OM that are relevant to their duties. Naturally, the operator should ensure that the authority is furnished with all amendments as they are issued. It is not necessary for the OM to contain information which is already available in the Aeroplane Flight Manual (AFM), which is part of the C of A.

Training Manual – ANO 29

For every UK-registered aircraft flying for public transport, the operator must ensure that a training manual is kept up-to-date and made available to every person appointed to give or supervise training, experience, practice or periodical tests as required by the regulations (see the previous chapter). As noted in the previous paragraph JAR-OPS regards this as an integral part of the OM.

Public Transport Operator's Responsibilities – ANO 30 and JAR-OPS 1.005 on

The operator shall not permit an aircraft to fly without first:

- designating a pilot to be the aircraft commander
- checking that all radio and navigation aids for the route and any possible diversion are adequate for safe navigation
- checking that all aerodromes that may possibly be used are properly manned and equipped to ensure the safety of the aircraft and its passengers
- ensuring that anyone carrying out crew duties is licensed, competent and able to use the aircraft equipment. The exception is when crew training is being carried out
- ensuring that no simulated emergency procedures are carried out which could adversely affect the flight characteristics of the aircraft

- ensuring that all employees are aware of all laws, regulations and procedures of those states in which the operations are being conducted
- ensuring that all crew members can communicate in a common language and should be able to understand the language used in the OM sections which are relevant to their duties
- ensuring that there is an approved MEL which is based upon the master minimum equipment list accepted by the authority (CAP 549). The aircraft must not be operated other than in accordance with this MEL without the authority's permission.

The operator is also responsible for:

- establishing a quality control system to ensure compliance with and the adequacy of procedures required to ensure safe operational practices and airworthy aeroplanes
- establishing an accident prevention and flight safety programme including programmes to achieve and maintain risk awareness by all persons involved in operations and evaluation of relevant information relating to accidents and incidents and the promulgation of related information
- ensuring that crew members not required for the flight have also been trained in and are proficient to perform their assigned duties
- ensuring that essential information relating to the intended flight regarding search and rescue services is accessible on the flight deck
- ensuring that an aeroplane with an approved seating capacity for over 30 passengers when flying over water is never more than 2 hours away at cruising speed or 400 nm from where an emergency landing can be made, whichever is the smaller, unless it is equipped with the correct ditching equipment
- ensuring that weapons and munitions of war are not carried unless the approval of all states concerned has been given. If carried they must be inaccessible to passengers and the weapons must be unloaded. The states concerned may agree to a variation in these rules. The commander must be given full details before a flight in all these cases
- ensuring that all sporting weapons and ammunition are subject to similar conditions. Under certain conditions ammunition for sporting weapons may be carried in passengers checked luggage
- ensuring that all reasonable measures are taken so that no person is carried in a part of the aircraft not designed for the purpose unless temporary access has been granted by the commander for the purpose of taking action necessary for the safety of the aeroplane or of any person, animal or goods
- taking all reasonable steps to ensure that no person offers or accepts

dangerous goods unless properly trained and the goods are correctly classified, documented, certificated, described, labelled and in a fit condition for transport by air.

Mass and Balance Documentation – JAR-OPS 1.625 and ANO 31

An operator shall establish mass and balance documentation prior to each flight, specifying the load and its distribution. The documentation must enable the commander to see that the mass and balance limits of the aeroplane are not exceeded. The person preparing the documentation must be named and the load supervisor must sign to indicate that the instructions have been followed. The commander will sign to indicate his acceptance of the loading. The operator must set up a procedure for dealing with Last Minute Changes (LMC). The OM must show the maximum permissible LMC and, if these are exceeded, new documentation must be prepared.

Computerised systems for dealing with mass and balance are acceptable but care must be taken to verify the integrity of the computer output. The output data should be positively verified every 6 months. If an operator wishes to use an on-board computer system as the primary source for these purposes, the authority must give its approval. When the documentation is sent to the aeroplanes via datalink, a copy of the documentation must be available on the ground.

One copy of the load sheet will be carried in the aircraft and one retained on the ground for 6 months.

Mass Values for Passengers and Baggage – JAR-OPS 1.620

An operator shall compute the mass of passengers and checked baggage using either the actual weighed mass or the standard mass given in tables in JAR-OPS. The tables take account of the number of passenger seats, whether males, females or children and whether the flight is a holiday charter flight or not. The standard masses for checked baggage pieces can only be used in aircraft with over 20 passenger seats and is dependent on whether the flight is domestic, within the European region or intercontinental. If the operator wishes to use his own values, he may do so after carrying out an elaborate sampling procedure and submitting the proposals to the authority. In any case, if it is suspected that the standard masses normally used are going to be incorrect, operators must carry out individual weighings or add an adequate mass increment and the commander must be advised.

Public Transport Operating Conditions – ANO 32 and JAR-OPS 1.240 on

The operator is responsible for ensuring:

- noise abatement procedures are complied with
- that the routes are safe considering:
  - ground facilities and services available
  - the aircraft equipment is adequate
  - appropriate documentation is available
  - all restrictions imposed by the authorities
- the aircraft has the correct equipment if it is to operate in Reduced Vertical Separation Minima (RVSM) areas (see Chapter 3)
- that the aircraft is correctly equipped when operating in Minimum Navigation Performance Specification (MNPS) areas (see Chapter 7)
- that, if the twin-engined aircraft has Extended Twin-jet Operations (ETOPS) approval, it remains within the quoted threshold distance/ time from an approved alternative
- that, if the twin-engined aircraft does not have ETOPS approval, routes will not be flown so that the aircraft is:
  - if it is in performance class A and with passenger seating of 20 or more or with a MSTOM of 45 360 kg, not more than 60 min at the one-engined cruise speed from an adequate aerodrome
  - if it is in performance class B or C, not more than 120 min at the one-engined cruise speed or 300 nm, whichever is the smaller, from an adequate aerodrome
- procedures are established for finding the Minimum Safe Altitudes (MSA) for all routes
- a fuel policy is established to make certain that every flight carries adequate fuel for all foreseeable contingencies
- a procedure is established for the carriage of Person with Reduced Mobility (PRM) and to check that they do not occupy seats where they impede the crew or obstruct access to emergency equipment or exits. The commander must be notified when PRMs are on board
- a procedure is established for the transportation of inadmissible passengers, deportees or persons in custody to ensure the safety of the aeroplane and its occupants. The commander must be notified when such persons are aboard
- procedures are established to ensure that baggage taken into the cabin can be adequately and securely stowed. All baggage and cargo on board that might cause injury or damage or obstruct the aisles or exits, if displaced, is placed in stowages to prevent movement
- procedures should ensure that passengers are seated where, in the

event of an emergency evacuation being required, they may best assist and not hinder evacuation.

Helicopters Flying Over Water – ANO 32 (5)

A helicopter with a performance classification B shall not fly over water for public transport for more than 20 seconds from whence it can make an autorotative descent to land in an emergency unless equipped with approved flotation gear and then only for 3 min unless the authority has given permission otherwise. If the performance classification is A, it shall not fly over water for more than 15 min unless equipped with approved flotation gear.

Passenger Briefing – JAR-OPS 1.285 and ANO 39

An operator is responsible for ensuring that passengers are briefed as follows:

- *Verbal briefing*
  about safety matters. Parts or all of this may be by an audio-visual presentation.
- *Briefing card*
  is issued on which picture type instructions indicate the operation of emergency equipment and the exits likely to be used.
- *Before take-off*
  the following must be dealt with, if applicable:
  - smoking regulations
  - seats to be upright and trays stowed
  - location of emergency exits
  - use of floor proximity escape path markings
  - stowage of hand baggage
  - restriction on use of electronic devices including telephones
  - location and contents of briefing card.
- *Demonstrations to be given*
  where applicable:
  - use of safety belts
  - use of oxygen equipment
  - location and use of life jackets.
- *In flight*
  when necessary remind passengers of:
  - smoking regulations
  - use of seat belts.

- *Pre-landing*
  remind passengers of:
  - smoking regulations
  - use of safety belts
  - seats to be upright and tables stowed
  - re-stowage of hand baggage
  - use of electronic devices.
- *After landing*
  remind passengers of:
  - smoking regulations
  - use of safety belts.

Aerodrome Operating Minima (AOM) – ANO 33–35, JAR 1.297, 1.400

All operators of aircraft using UK aerodromes shall establish AOM considering:

- *The aircraft*
  type, performance and handling characteristics and any relevant conditions in its C of A.
- *Composition of crew.*
- *The relevant aerodrome*
  physical characteristics and its surroundings.
- *Runway dimensions.*
- *Landing aids available.*

An aircraft shall not commence a flight at a time when:

- the cloud ceiling or the runway visual range (RVR) at the departure aerodrome is less than the specified minima for take-off
- considering the landing AOM for the destination and its alternative, a landing would not be practicable.

An aircraft, when making a descent to an aerodrome, shall not descend below 1000 ft above the aerodrome (or the outer marker) if any of the RVR being broadcast for the runway are less than the specified minima. In addition, an aircraft shall not:

- continue an approach to landing to below the specified decision height (DH) when using a precision aid such as full ILS
- descend below the specified Minimum Descent Height (MDH) when using a non-precision approach aid such as ILS without a glide path

unless the specified visual reference for landing has been established and maintained.

The following definitions should be noted:

- *Specified*
  as laid down in the OM or if there is no OM, as in the case of private aircraft, it refers to the published AOM for the aerodrome or calculated by a notified method.
- *Take-off AOM*
  cloud ceiling and RVR.
- *Landing AOM*
  DH or MDH and RVR.
- *Cloud ceiling*
  the vertical distance from the aerodrome to the lowest part of any cloud visible from the aerodrome obscuring more than half the sky.

Crew Members at Stations – JAR-OPS 1.310 on, ANO 36

- *Flight crew*
  - during take-off and landing each flight crew member required to be on the flight deck is on station
  - during all other phases, flight crew required to be on flight deck duty remain on station unless absence is necessary for the operation or for physiological needs provided that there is always one suitably qualified pilot at the controls.
- *Cabin crew members*
  to be seated at their assigned stations during take-off and landing and when thought necessary by the commander in the interests of safety.

Seats, Safety Belts and Harnesses – JAR-OPS 1.320 and ANO 37

- *Crew members*
  during take-off and landing and whenever the commander thinks it necessary, each crew member should be properly secured by all safety belts and harnesses provided. At all times, crew members on the flight deck shall keep their safety belts fastened while at their station.
- *Passengers*
  - before take-off and landing, while taxying and whenever the commander considers it necessary, passengers must keep their seat belts fastened

– provision shall be made for, and the commander shall ensure that multiple occupancy of a seat only occurs when it is one adult and an infant properly secured by some special device.

Security and Accessibility to Exits – JAR-OPS 1.325 and 1.326

Procedures should be established to ensure that before taxying, take-off and landing, all exits and escape paths are unobstructed. The commander should ensure that before take-off and landing and whenever necessary, all equipment and baggage is properly secured. At all times, the commander should ensure that all emergency equipment remains readily available.

Smoking on Board ANO-OPS 1.340 and ANO 58

The commander shall ensure that no person is allowed to smoke:

* whenever the commander thinks it necessary and under UK law a non-smoking sign should be exhibited which is visible to all passengers
* while on the ground unless permitted under procedures set out in the OM
* outside designated smoking areas, in the aisles and in the toilets
* in cargo compartments and/or other areas where cargo is carried not protected by flame-proofing
* in areas where oxygen is being supplied.

Persons Endangering Aircraft and Other Persons – JAR-OPS 1.105, 1.120 and ANO55 and 56

Persons shall:

* not secrete themselves or cargo on board an aircraft
* not operate a portable electronic device that can adversely affect the aircraft systems
* not act so as to endanger the aeroplane and persons on board or cause the aeroplane to endanger persons or property
* not disobey the lawful commands given by the commander so as to secure the safety of the aeroplane, persons and property
* only be carried in proper accommodation, not towed in anything other than a glider or a flying machine

- only have temporary access to any part of the aircraft in the interest of safety or to that part of the aircraft used for carrying cargo and designed to have proper access.

Oxygen Equipment – JAR-0PS 1.760 and ANO 40

A public transport aircraft that is to be flown over 10 000 ft must have a supplemental oxygen supply. The amount of oxygen available must be determined considering the cabin pressure altitude, the flight duration and the assumption that pressurisation failure occurs at the most critical point from the point of oxygen need and that after failure the aeroplane will descend in accordance with the procedures specified in the Aeroplane Flight Manual (AFM) to a safe altitude for the route to be flown.

Quite elaborate rules are laid down in JAR-OPS for the minimum requirements of supplemental oxygen to be carried. It is particularly significant that supplies for all crew will be sufficient for the entire flight time when the cabin pressure altitude exceeds 13 000 ft. Aeroplanes to be operated above 25 000 ft or if operated below this cannot descend within 4 min to 13 000 ft and for which the C of A was first issued after 9 November 1998 shall be provided with automatically deployable oxygen equipment. In addition, if pressurised aeroplanes are to be operated over 25 000 ft, there should be a supply of pure oxygen in the passenger cabin for first-aid purposes.

Areas Requiring Special Equipment – ANO 42, 43, 44 and JAR-OPS 1.865

As already mentioned in Chapter 7, flight in certain areas and on certain routes will require special equipment:

- *Minimum Navigation Performance Specification (MNPS) areas.*
- *Routes Requiring Area Navigation (RNAV) equipment*
  these routes will be shown on route charts with 'R' preceding the route name.
- *Reduced Vertical Separation Minima (RVSM) areas*
  mostly the large ocean areas such as the North Atlantic.

Flight Recorders – JAR-OPS 1.160, 1.700, 1.715 and ANO 45

In general, public transport aircraft that are turbine powered and with a MSTOM over 5700 kg or with passenger seating over nine are required to carry Flight Data Recorders (FDR) and Cockpit Voice Recorders (CVR). There are various specifications for the equipment depending on the date the C of A was issued and the actual date of the operation. In some cases it will refer to any turbine-powered aircraft with a MSTOM over 5700 kg, regardless of the passenger seating.

Where required, FDR and CVR must be in use from the start of the take-off run to the end of the landing run or in the case of helicopters for all the time that the rotors are turning. Unless the authority says otherwise, following an incident that is subject to mandatory reporting, the records should be preserved for 60 days or as long as the investigating authority requires them. In any case, flight recordings should always be preserved for 60 days unless otherwise instructed by an investigating authority.

Towing of Gliders – ANO 46

Under UK law the following conditions apply:

- *Certificate of airworthiness*
  must specifically state that this is authorised.
- *Total length of aircraft, rope and glider*
  must not exceed 150 m.
- *Commander's pre-take-off action*
  - check everything is satisfactory and all should be safe until the intended landing
  - signals have been agreed between everybody to ensure a safe take-off
  - emergency signals have been agreed between the two pilots so that the aeroplane commander can suggest immediate release and the glider pilot can indicate that this is not possible.

Towing, Picking Up and Raising of Articles and Persons – ANO 47

Under UK law, unless the C of A specifically permits, an aircraft is not permitted to tow any article other than a glider or pick up or raise any person, animal or article. These procedures are not prohibited in an emergency or for the purpose of saving life. Apart from gliders, any towing must be by day and the visibility must not be under 1 nm. An aircraft must only launch or pick up tow ropes, banners or similar articles at an aerodrome. A helicopter shall not fly over a congested area when an article, person or animal is suspended from it and shall not carry any passengers unless they have duties in connection with the load.

Dropping of Animals and Articles – ANO 48

Under UK law, articles and animals are not to be dropped so as to endanger persons and property. Over the UK, articles and animals are not to be dropped even when attached to a parachute except:

- to save life
- jettisoning of fuel and other articles in an emergency
- dropping of ballast in the form of fine sand and water
- dropping articles for navigation purposes
- dropping of ropes, banners or similar articles at an aerodrome
- for public health purposes or as a measure against weather conditions, surface icing or oil pollution or training for these purposes
- dropping of wind indicators when parachutists are about to be dropped.

Permission can be obtained for exemption from this regulation or for parachuting. In addition, aerial application certificates may be granted for agricultural and forestry purposes and when operating in this role, the minimum 500-ft level (see low flying regulations in Chapter 19) does not apply.

Parachute Descents – ANO 49

Parachuting, except in an emergency, is only permitted over the UK subject to the following conditions:

- in accordance with CAA permission or a police AOC
- persons and property must not be endangered
- the C of A must specifically allow it
- the operator must provide the CAA with his manual and everyone concerned must have access to it.

Aircraft Security Requirements JAR-OPS 1.1235 onwards

It is the operator's responsibility to ensure that:

- all appropriate personnel comply with the national security programmes
- an approved training programme is in operation to enable personnel to take action to prevent acts of unlawful interference such as sabotage or unlawful seizure of aeroplanes

- all acts of unlawful interference are immediately reported to the authorities
- all aircraft carry a checklist of the search procedure to be followed when looking for weapons, explosives or other dangerous devices
- all flight crew compartment doors, if installed, should be capable of being locked.

Emergency Equipment to be Carried JAR-OPS 1.830 onwards

An operator shall not operate a public transport aeroplane unless carrying:

- *Automatic Emergency Locator Transmitter (ELT)*
  should transmit in the event of a crash but the transmitting at other times is minimised.
- *Life jackets equipped with lights*
  when flying over water more than 50 nm off land or where the take-off or approach path is so disposed that, in the event of a mishap, ditching is possible. Seaplanes and amphibians must always carry life jackets.
- *Life-rafts and survival ELTs*
  on overwater flights at a distance, from where an emergency landing could be made, greater than 120 min at cruising speed or 400 nm whichever is the lesser, for aeroplanes capable of continuing flight with the critical power unit(s) becoming inoperative at any point or 30 min or 100 nm, whichever is the lesser for all other aeroplanes.
- *Survival equipment*
  on any flight where search and rescue would be difficult unless equipped with the following:
  - pyrotechnic signalling equipment
  - at least one ELT
  - additional survival equipment taking into account the number of persons aboard.
- *Sea anchor and sound signalling equipment*
  for all seaplanes and amphibians.

Crew Interphone System – JAR-OPS 1.690

An operator must not operate a public transport aeroplane with a MSTOM 15 000 kg or having maximum passenger seating of over 19 unless it is equipped with a crew interphone system. This is the general rule but there are exceptions for older aircraft first given a C of A before 1 April 1965.

Public Address System – JAR-OPS 1.695

An operator of a public transport aeroplane with maximum seating for passengers of over 19 must have an approved public address system. It should:

* operate independently of the interphone system
* be readily accessible at every flight crew member station
* where emergency exits have cabin crew seats next to them, have a convenient microphone
* be capable of operation within 10 seconds by each cabin crew member
* be audible and intelligible everywhere passengers are likely to be

Altitude Alerting System JAR-OPS 1.660

An operator shall not operate a turbine propeller aeroplane with a MSTOM over 5700 kg or with passenger seating of over nine or any turbojet unless it has an altitude alerting system which alerts flight crew of the approach to a pre-selected altitude and also when deviating from a pre-selected altitude. Smaller aircraft with MSTOM of 5700 kg or less but seating more than nine passengers and first issued with a C of A before 1 April 1972 are exempted.

Ground Proximity Warning System (GPWS) – JAR-OPS 1.665

A GPWS must automatically provide aural and possibly visual signals giving timely and distinctive warning to the flight crew of sink rate, ground proximity, altitude loss after take-off or go-around, incorrect landing configuration and downward slope deviation. It is required in all public transport turbine-powered aeroplanes having a MSTOM over 15 000 kg or a maximum approved passenger seating of over 30 or after 1 January 1999 having a MSTOM of over 5700 kg or maximum passenger seating of over nine.

Airborne Weather Radar (AWR) – JAR-OPS 1.670

An operator shall not operate a public transport pressurised aeroplane or an unpressurised one with a MSTOM over 5700 kg or, after 1 April 1999, having more than nine passenger seats unless it has AWR whenever the aeroplane is operated at night or in Instrument Meteorological Conditions (IMC) where thunderstorms or other potentially hazardous conditions which would be detectable, may be expected.

For public transport propeller-driven pressurised aircraft having a MSTOM not over 5700 kg and with passenger seating of nine or less, the AWR may be replaced by other equipment, approved by the authority, capable of detecting thunderstorms and other hazardous weather conditions.

Airborne Collision Avoidance Systems (ACAS) – JAR-OPS 1.668

Many public transport aircraft are already carrying this equipment to comply with the regulations in other countries. Under JAR-OPS it will be a requirement after 1 January 2000 for turbine-powered aircraft having a MSTOM over 15 000 kg or a maximum passenger capacity of over 30 to be fitted with equipment having a performance of at least that specified for ACAS II in ICAO Annex 10 Volume 4 – 'Surveillance Radar and Collision Avoidance Systems'. After 1 January 2005, this will be extended to MSTOM of over 5700 kg or a maximum passenger seating capacity of over 19. A yellow AIC gives more details.

Chapter 19
Documents and Reports

Introduction

Many references to documents have already been made. In this chapter, brief references will be made to these and other documents will be described. In addition, the documents to be carried on the aircraft and retained by the operator will be dealt with.

Documents to be Carried – JAR-OPS 1.125 and ANO 66, Schedule 11

A public transport operator shall ensure that the following documents or copies are carried on each flight:

- certificate of registration
- certificate of airworthiness
- noise certificate – if applicable
- air operator's certificate
- aircraft radio licence
- third-party liability insurance certificate
- valid crew licences (carried by individuals)
- *Operations Manual (OM) – sections required in flight
- aeroplane flight manual unless all necessary information is in OM
- *operational flight plan
- *aeroplane technical log supported by Certificate of Maintenance Review
- *ATS flight plan
- *appropriate NOTAMs/AIS briefing documents
- *appropriate meteorological information
- *mass and balance documentation
- *details of special categories of passengers such as security personnel, handicapped persons, inadmissible passengers, deportees and persons in custody
- *special loads notification including dangerous goods and the information given to the commander

- *current maps and charts including approach and aerodrome charts for the operation
- *any other documents required by the states concerned such as cargo and passenger manifests, general declarations, etc.
- *forms to comply with the reporting requirements of the authority and the operator.

The authority may permit the information marked with asterisks (*) to be presented in a form other than on printed paper.

International Private Flight – ANO Schedule 11

In this case only the following need be carried:

- certificate of airworthiness
- certificate of registration
- radio licences
- crew licences
- interception procedures (for public transport these will be included in the OM).

Information Retained on the Ground – JAR-OPS 1.140

An operator shall retain, for the duration of each flight or series of flights, information relevant to the flight and appropriate to the type of operation. The information is retained until it has been duplicated at the place at which it will be stored. The information referred to is:

- copies of the relevant parts of the technical log
- a copy of the operational flight plan
- route specific NOTAM documentation if specifically edited by the operator
- mass and balance documentation, if required
- special loads documentation.

Apart from the first item (technical log) which should be retained for 2 years from the date of the last entry, all the others should be kept for periods of three months. The operator is responsible for preserving this documentation even if he ceases to be the operator of the aeroplane.

Inspection and Production of Records – JAR-OPS 1.145 on, ANO 68 onwards

An operator shall permit any authorised person to inspect any documents or records related to flight operations or maintenance and, on request, produce all such records to the authority within a reasonable time. The commander shall within a reasonable time produce to the authority the documentation required to be carried on board.

Times Other Records and Reports need to be Kept – JAR-OPS 1.1065

- *Reports*
  the following to be kept for 3 months: journey logs, flight reports, when considered necessary, reports on exceedance of duty and/or reducing rest periods.
- *Flight crew records*
  the following to be kept for 3 years:
  - conversion training and checking
  - command courses and checking
  - recurrent training and checking to operate in either of the pilot's seats
  - route and aerodrome competence
  - specific operations training and qualification (e.g. ETOPS or Cat II/III operations)
  - dangerous goods training
  the following to be kept for 15 months:
  - flight, duty and rest time
  - recent experience of take-offs and landings.
  Licences will be kept as long as the crew member is exercising the licence privileges.
- *Cabin crew records*
  - flight, duty and rest time will be kept for 15 months
  - initial training, conversion and differences training for as long as in employment
  - recurrent training and refresher training for 12 months after leaving employ
  - dangerous goods training, where appropriate, for 3 years.
- *Other records*
  - the last two training records for other operations personnel where JAR-OPS requires training

– cosmic and solar radiation dosage to be kept for 12 months after the crew member has left the employ
– quality system records to be kept for 5 years.

Occurrence Reporting – JAR-OPS 1.420 and ANO 106, AN(G)R 17

- *Flight incidents*
  the operator or commander shall report to the authority, within 72 hours, any incident that has endangered or may have endangered a flight.
- *Technical defects and exceedance of technical limits*
  the commander should record all these in the technical log.
- *Air traffic incidents*
  the commander shall report in accordance with ICAO PANS RAC whenever an aeroplane in flight has been endangered by:
  – near air collision (see AIRPROX, Chapter 11).
  – faulty air traffic procedures or lack of compliance with procedures by ATS or flight crew.
  – a failure of ATS.
- *Bird hazards and strikes*
  a commander shall immediately inform the appropriate ground station when a bird hazard is observed and if there is a bird strike a written report should be submitted on landing.
- *In-flight emergencies with dangerous goods aboard*
  when the situation permits, the commander should inform the ATS unit of any dangerous goods aboard.
- *Unlawful interference*
  following such an act the commander shall submit a report to the appropriate authorities as soon as possible after landing.
- *Irregularities of ground and navigational facilities and hazardous conditions*
  when encountered in flight, a commander should notify the appropriate ground station as soon as possible of potentially hazardous conditions such as:
  – an irregularity in a ground or navigational facility
  – a meteorological phenomenon
  – a volcanic ash cloud
  – high radiation level.

Accident Reporting – JAR-OPS 1.425 and Civil Aviation (Investigation of Accidents) Regulations

JAR-OPS merely state that in the event of an accident the nearest appropriate authority should be notified, by the quickest possible means, of any accident resulting in serious injury (as defined in ICAO Annex 13) or death of any person or substantial damage to the aeroplane or property. Under UK law an accident is 'reportable' to the Air Accidents Investigation Branch (AAIB) if it occurs between the time when any person boards the aircraft with the intention of flight until such time as all persons have disembarked. The types of accidents that are reportable are when:

- *Persons*
  suffer death or serious injury while in or upon the aircraft or by direct contact with any part of the aircraft or any part which has become detached from the aircraft or by direct exposure to jet blast.
- *Aircraft*
  suffers damage or structural failure which would adversely affect the aircraft's structural strength, performance or flight characteristics and would normally require major repair or replacement other than:
  – engine failure or damage, when the damage is limited to the engine, its cowling or accessories
  – damage to the propellers, wing-tips, antennae, tyres, brakes, fairings, small dents or puncture holes in the aircraft skin.
- *Aircraft is missing or completely inaccessible.*

What is meant by serious injury is given in detail in the accident regulations. Death or serious injury from natural causes, or if self-inflicted or if suffered by a stowaway hiding in areas not normally available in flight to passengers or crew, are not reportable. Accidents will also be reportable to the local police if there is serious injury, death or damage to property.

Air Operator's Certificate (AOC) – JAR-OPS 1.175 and ANO 6

Aircraft flying for public transport must have an AOC. Before issuing, the authority will need to be satisfied with the operator, the operator's organisation, staff, maintenance arrangements and capability to secure the safe operation of the approved aircraft in approved areas. An essential part of the procedure for obtaining an AOC, is to produce and get approved an Operations Manual (OM). Once approved this becomes the 'bible' for the operator and staff. Everything must be carried out in accordance with the

OM and it is an offence under the law not to do so. Periodical inspections will be carried out by the authority's staff to check that this is being done. Inspections will be made of the maintenance organisation and the operation of the aircraft. The authority's inspectors will have the right of access to all records and are entitled to fly on the operator's aircraft either on normal operations or when flight crew testing is being carried out. Inspectors may have access to the flight deck, but the commander may refuse this access if he or she considers that the safety of the operation or the test would be affected.

Interception Procedures – ANO Schedule 11

JAR-OPS 1 only makes a passing reference to these when listing the contents of the OM. It refers to interception procedures under the general heading of rules of the air. Under UK law it is a specific requirement that all aircraft operating internationally should carry a document giving the agreed procedures of the actions to take when intercepted by military aircraft. This usually occurs if straying too close to military sensitive areas. Carrying this document as part of the OM will obviously satisfy this requirement.

Journey Log – JAR-OPS 1.1055 and 1.415

It is a requirement that the flight commander must complete and sign a journey log. This will be retained by the operator and should contain:

• aeroplane registration
• date
• crew members and their duties
• route details, departure and arrival aerodromes and on and off block times
• nature of flight
• incidents and observations (if any).

If the information is available elsewhere, an operator may be excused from having a journey log kept.

Operational Flight Plan – JAR-OPS 1.060

The operator must ensure that the operational flight plan used and the entries made during flight contain the following items:

- *Flight details*
  registration and flight identification, aeroplane type, date.
- *Crewing*
  crew members and their duties.
- *Route details*
  departure and arrival points, off and on blocks times. Type of operation – ETOPs, Ferry flight, VFR, etc. Routing with check-points, distances and tracks.
- *Speeds and times*
  planned cruising speed and estimated and actual times over check-points.
- *Altitudes and flight levels*
  safe altitudes and minimum levels. Planned altitudes and flight levels.
- *Fuel*
  fuel calculations and records of in-flight checks. Fuel on board when starting engines.
- *Alternates for destination and departure (where necessary)*
  full details including all details as in the last four items above.
- *ATS flight plan and clearance.*
- *In-flight*
  any re-planning calculations and relevant meteorological information.

If the information is available elsewhere, or is irrelevant to the particular operation, it may be omitted from the operational flight plan. The operational flight plan procedure must be described in the OM. All entries should be made concurrently and should be permanent in nature. The storage period is 3 months.

Personal Flying Log Books – ANO 24

JAR-OPS make no reference to these, but obviously every pilot will need to keep one as a record of their own experience which will be required when applying for licences. The ANO deals with them in some detail and the following points are made:

- it shall include the name and address of the pilot and his employer (if any) together with licence particulars

- full details of each flight made should be entered, as soon as possible, including not only the times but the aircraft used and the capacity in which the holder acted on the flight
- special conditions on the flight including night and instrument flying and any test taken on the flight
- details of any test taken in a flight simulator shall be recorded
- it shall be produced within a reasonable time to any authorised person after being requested to do so within 2 years of the last entry being made.

Offences in Relation to Documents and Records – ANO 73

A person shall not with intent to deceive:

- *Use forged, altered, revoked, suspended or non-applicable documents.*
- *Lend licences, certificates, etc.*
  to any other person.
- *Procure licences or any other documents*
  by false representation.
- *Mutilate, alter or render illegible any document.*
- *Make a false entry or make a material omission or destroy a document*
  all entries in log books or records should be made in ink or indelible pencil.

Chapter 20
Rules of the Air, Aircraft Lights and Marshalling Signals

Introduction

All countries in ICAO will follow the Standard Rules and Recommended Procedures (SARPs) of Annex 2 – Rules of the Air. The UK document in question is 'The Rules of the Air Regulations' which are supplemented by the Civil Air Publication (CAP 637).

Authority for Rules of the Air (R of A) – ANO 74

The ANO authorises the making of regulations prescribing:

- how an aircraft may move including provision to give way to military aircraft
- the lights and other signals to be shown or made by aircraft or persons
- the lighting and marking of aerodromes
- any other provisions for securing the safety of aircraft in flight and in movement and the safety of persons and property on the surface.

Departing from the Rules of the Air – ANO 74

It is usually an offence under UK law to depart from the R of A but it is permitted:

- for avoiding immediate danger
- for complying with the law of another country in which the UK aircraft happens to be
- for complying with military flying regulations.

If any departure is made from the R of A for the first of these reasons, a report should be made by the commander to the appropriate authority.

Nothing in the R of A shall exonerate a person from the consequences of any neglect in the use of lights or signals or of the neglect of precautions required by normal aviation practice.

Rules for Avoiding Aerial Collision – R of A 17 (1)

Even if a flight is made with ATC clearance, it is the duty of the commander to take all possible measures to avoid collision with another aircraft. The following general rules should be observed:

- do not fly dangerously close to another aircraft
- do not formate with other aircraft unless all commanders have agreed to do so
- if obliged to give way under R of A, avoid passing over, under or ahead of the other aircraft unless well clear
- an aircraft which has the right of way, shall maintain its course and speed
- a glider and the towing aircraft are considered to be a single aircraft
- aircraft flying under a police Aircraft Operator's Certificate (AOC) are not subject to the rule about formatting and maintaining course and speed.

Rules for Converging Aircraft – ANO 17 (2)

- *Order of priority*
  converging aircraft are classified as:
  - flying machines
  - flying machines towing anything
  - airships
  - gliders
  - balloons.

  Each type of aircraft must give way to all aircraft below it on the list.
- *General rule*
  when two aircraft are converging in the air at approximately the same altitude, the aircraft, which has the other on its right, shall give way. In other words the aircraft on the right has the right of way (Figure 20.1).

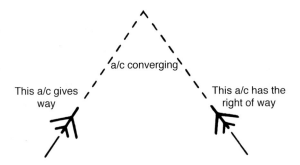

Figure 20.1 Right of way for converging aircraft.

Rules for Aircraft Approaching Head-on and Overtaking – ANO 17 (3) and (4)

- *Approaching head-on*
  if there is danger of a collision, each aircraft alters heading to the right.
- *Overtaking*
  gliders may alter course in either direction but with powered aircraft the one in front has the right of way and the overtaking aircraft must keep out of the way by altering course to the right and shall not cease to keep out of the way until the other aircraft has been passed and it is well clear.
- *Flight in the vicinity of an aerodrome*
  a flying machine, glider or airship while flying here or moving on an aerodrome shall, unless ATC says otherwise, conform to the pattern of aircraft intending to land or keep clear of the airspace in which the pattern is formed. Unless notified or signalled otherwise all turns will be made to the left.
- *Order of landing*
  these rules have already been mentioned in Chapter 8, but are repeated here for the sake of clarity:
  - an aircraft while landing or on the final approach has priority over all other traffic. If two or more flying machines, gliders or airships are approaching at the same time, the aircraft at the lower altitude has the right of way but it shall not cut in front of or overtake another which is on its final approach
  - if the ATC has given an order of priority, this must be obeyed
  - if the commander is aware that another aircraft is making an emergency landing, then right of way should be given to that aircraft and, if at night, shall await permission to land.

- *Landing and taking-off*
  a flying machine, glider or airship shall take-off and land in the direction shown by ground signals or, failing these, into wind unless good aviation practice demands otherwise. A flying machine or glider shall not land on a runway unless it is clear of other aircraft or unless ATC authorises it. Where take-off and landings are not confined to a runway:
  - a flying machine or glider, when landing, shall leave clear on its left any aircraft that has landed or has already landed or is about to take off. On landing the aircraft shall turn left, having checked that it is safe to do so
  - a flying machine taking off shall leave clear on its left any other machines that are taking off
  - on landing flying machines shall move clear of the landing area as soon as possible unless ATC says otherwise.

Rights of Way on the Ground – ANO 37

These rules apply to flying machines on any part of a land aerodrome provided for the use of aircraft and under control of the person in charge of the aerodrome. In spite of any ATC clearance, it remains the duty of the commander to take all possible steps to avoid collisions with other aircraft or vehicles. In the following list each has priority over the following aircraft and/or vehicles:

- taking off and landing aircraft
- vehicles towing aircraft
- aircraft
- vehicles.

The following are the rights of way rules on the ground:

- *Aircraft approaching head-on*
  each alters heading to the right.
- *Aircraft converging*
  the one on the right has the right of way.
- *Overtaking aircraft*
  the aircraft being overtaken has the right of way and the overtaking aircraft should alter course to the left until the other aircraft has been passed and is well clear.

Notice the agreement with the rules that apply in flight for the first and second of these rules. For ground vehicles at UK aerodromes the normal rules of the road apply – keep to the left and overtake on the right.

Aerobatic Manoeuvres – R of A 18

These are not permitted over the congested areas of any city, town or settlement. Within Controlled Airspace (CAS) they may only be carried out with ATC approval.

Right-hand Traffic Rule – R of A 19

An aircraft flying within sight of the ground and following a road, railway or similar line feature shall keep the landmark on its left. The aircraft will be flying on the right following the continental rule of the road. The rule will not apply in CAS where the ATC instructions must be obeyed.

Simulated Instrument Flight – R of A 6

An aircraft may only be flown in simulated instrument flight conditions if:

- *Dual controls are fitted.*
- *A safety pilot*
  is carried in the second control seat who can render assistance if necessary.
- *A competent observer*
  must be carried if the safety pilot's field of vision is not adequate.

Practice Instrument Approaches – R of A 7

These may only be carried out in Visual Meteorological conditions (VMC) if:

- the ATC has been warned in advance
- when the flight is not being carried out in simulated instrument flight conditions, a competent observer is carried.

Display of Lights by Aircraft – ANO 8 –11

Lights and other signals to be shown by aircraft are listed in Table 20.1.

Table 20.1 Lights and other signals to be shown by aircraft

Type of aircraft	Situation or state	Lights to display	Notes including R of A rule number
Any	Stationary on apron or in maintenance area	None	9 (2) a
Any	As above with engines running	ACL	9 (2) b
Any	Moving on aerodrome	Same as for flight or BL steady or flashing	9 (2) a
Any fitted with ACL	In flight day or night	ACL	9(1) b
UK-registered MSTOM over 5700 kg or any mass if registered after April 1988	In flight at night	BL + ACL	11(1) a
UK-registered before April 1988 and 5700 kg or less	In flight at night	BL ± ACL or BL flashing ± ACL	11(1) b
Non-UK-registered.	In flight at night	BL ± ACL or BL flashing ± ACL or BL alternating with OD white ± ACL or BL alternating with red tail light ± ACL or combination of the last two	11(1) c
Gliders	In flight at night	Steady OD red or as in R of A 11(1) a or b	12
Free balloons	In flight at night	Steady OD red	13
Captive balloons or kites	In flight at night	White over red below basket both OD. On mooring cable similar lights at 300 m intervals	14 (1). Mooring marked with triangle of two red and one green light
Captive balloons or kites	In flight by day	Tubular red and white streamers at 200 m intervals	14 (2) and (3)

(table continues overleaf)

Table 20.1 *Continued*

Type of aircraft	Situation or state	Lights to display	Notes including R of A rule number
Airships	In flight at night	BL + ACL and white light showing forward 110° either side of dead ahead	15 (1)
Airships	In flight at night, not under command or with engines stopped or is being towed	White light showing forward 110° either side of dead ahead and white light showing to rear 70° to either side plus two red lights one above the other beneath the airship	15 (2)
Airships	Moored at night on mooring mast	White light OD near to rear	15 (3) a

BL, basic aircraft lights (see Figures 20.2 and 20.3); ACL, anti-collision light; OD, omni-directional.

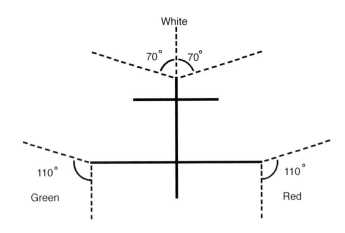

Figure 20.2 Plan view of basic aircraft lights.

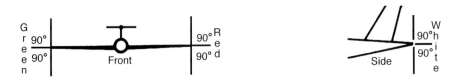

Figure 20.3 Basic aircraft lights.

Anti-collision Lights – R of A 1

These are lights showing in all directions so that the aircraft may be readily detectable at a distance. They may take the form of:

- in the case of a rotorcraft, a flashing red light
- for other aircraft, a flashing red or white light.

Failure of Navigation Lights – R of A 10

If this occurs on the ground, the aircraft shall not depart. If in flight, the aircraft should land as soon as it is safe to do so unless authorised by the ATC to continue the flight. If an anti-collision light fails by day, the flight may continue, provided the light is replaced as soon as possible.

Avoiding Collisions at Night – R of A 17

These rules are exactly the same at night as by day, but the pilot has to interpret them by the lights that are visible and their relative movement. If an aircraft light is seen at approximately the same height and there is no apparent movement, that is the bearing remains reasonably constant, and it appears to be getting closer, a collision risks exists. It will now have to be decided whether you or the other aircraft has the right of way. The difficult cases will be when the aircraft are converging and the following doggerel rhyme may prove a useful memory aid:

Red indicates port side (left) and green starboard side (right)

Red on Red or Green on Green

Indicates that all's serene

But, colours of a different hue

Indicates some action's due

If on your Red, you see a Green

Theirs should be the action seen

But, if on your Green you see a Red

YOU'D BETTER ACT OR YOU'LL BE DEAD

Marshalling Signals from Marshaller to a Pilot – R of A 47

Descriptions are given in Table 20.2.

Table 20.2

Description of signal	Meaning	Signal
Right or left arm down, other arm moves across the body and extended to indicate position of next marshaller	Proceed under guidance of next marshaller	
Arms moved repeatedly upwards and backwards, beckoning onwards	Move ahead	
Right arm down, left arm repeatedly moved upward and backward. The speed of movement indicates the rate of turn	Open up starboard engine or turn to port. Opposite arm indicates the opposite engine and direction of turn	
Arms crossed repeatedly above the head. The speed indicates the degree of urgency	Stop	
A circular motion of the right hand at head level, with the left hand pointing to the appropriate engine	Start engine pointed to	

Arms extended, the palms facing inwards, then swung from the extended position inwards	Chocks inserted	
Arms down, the palms facing outwards, then swung outwards	Chocks away	
Either arm and hand placed level with the chest, then moved sidewards with the palm downwards	Cut engines	
Arms placed down, with the palms towards the ground, then moved up and down several times	Slow down	
Arms above the head in a vertical position	This bay	
Arms placed down, with the palms towards the ground, then an arm moved up and down	Slow down engine on the side indicated by the arm being moved	
The right arm raised at the elbow, with the hand facing forward	All clear; marshalling finished	

Signals from Pilot to Marshaller – R of A 48

Descriptions are given in Table 20.3.

Table 20.3

Meaning of signal	Description of pilot's signal
Brakes applied	Extend, arm, hand and fingers in front of face, then clench fist
Brakes released	Clench fist in front of face, then extend fingers
Insert chocks	Arms extended, palms outwards, then move hand inwards to cross in front of face
Remove chocks	Hands crossed in front of face, palms outwards, then move arms outward
Ready to start engine	Raise number of fingers on one hand to indicate the engine to be started. Engines numbered from the pilot's left to right. Port outer is 1, etc.

Chapter 21
Miscellaneous Information

Introduction

This chapter contains various items not considered large enough to merit complete chapters of their own.

Security

In Chapter 1 reference was made to the ICAO Annex 17 – 'Security – Safeguarding International Civil Aviation against Acts of Unlawful Interference' and to the Tokyo Convention which was concerned with security. The following terms have specific definitions in these documents:

- *Airside*
  the movement area of an airport, adjacent terrain and buildings or portions of buildings to which access is controlled.
- *Security*
  a combination of measures and human and material resources intended to safeguard international civil aviation against acts of unlawful interference.
- *Security control*
  a means by which the introduction of weapons, explosives or articles likely to be used to commit an act of unlawful interference can be prevented.
- *Security programme*
  measures adopted to safeguard international civil aviation against acts of unlawful interference.

The ICAO's general objective is to have every contracting state take adequate security measures and then to issue a security manual (Document 8973) giving guidance on this topic. In particular there is emphasis on the need to pay special attention to the threat posed by explosive devices concealed in, or using electric, electronic or battery-operated items carried as hand-baggage or in checked baggage.

Security organisation

ICAO recommend the following procedures:

- *Inspection*
  each state should ensure arrangements are made for the inspection of suspected sabotage devices or other potential hazards at international airports. Research should be promoted into development of new equipment to improve security objectives.
- *International co-operation*
  with other states is recommended and when any bilateral agreement on air transport is made, reference to security should be included.
- *Sky marshals*
  these are allowed by some states. Special authorisation should be given for the carriage of arms. The commander must be aware of their presence on board and their location.
- *Pre-flight checks on all international flights*
  these should always be made to include measures designed to discover weapons, explosives and sabotage or other devices.
- *Passengers subject to judicial or administrative procedures*
  operators to include in their security programmes to ensure safety on board when carrying these passengers.
- *Disembarking passengers*
  operators to ensure they do not leave items on board.
- *Unaccompanied baggage*
  operators to ensure:
  - no baggage transported unless the owner is on board, unless the baggage is subject to other security checks
  - consignments and other baggage to be carried on passenger flight and using commercial courier services are subject to special security control
  - baggage for passenger flights when originating from places other than the airport check-in counter is subject to security checks
  - storage areas where mishandled baggage may be held until dealt with
  - unidentified baggage is kept in an isolated area until security-checked.
- *Unauthorised personnel*
  ICAO recommend that identity systems are established to prevent these getting airside or into any other security-sensitive areas.
- *Access to aircraft*
  adequate supervision of persons to and from aircraft must be exercised and steps taken to ensure there is no unauthorised access to aircraft.

Instruction in Flying – ANO 25

A pilot shall not give flying instruction unless he or she has a valid licence to be pilot in command of the aircraft and for the circumstances of the instruction, together with a flying instructor's or assistant flying instructor's rating. The article applies to any flying instruction for the purpose of qualifying for a pilot's licence or the inclusion or variation of any rating to the licence. It does not apply to flights for the inclusion of a rating to act as pilot of a multi-engined aircraft if that person has had a previous entitlement or has been a service pilot of that class of aircraft.

Balloons, Kites, Airships, Gliders and Parascending Parachutes – ANO 76

Table 21.1 gives details of when various aircraft may be flown or launched. The abbreviation NATZ indicates a Notified Aerodrome Traffic Zone and NOH indicates Notified Operating Hours. Notified means notified in NOTAMs or in the UKAIP Aerodrome (AD) section.

Regulation of Small Aircraft – ANO 76 A

These are defined in ANO 118 (1) as pilotless aircraft not over 20 kg, without fuel but with all flight equipment. The person in charge of such an aircraft must not fly it:

- unless satisfied that the flight can be made safely
- in class A, B, C, D or E airspace without ATC permission
- within an ATZ during the notified hours of watch of the ATC without ATC permission
- above 400 ft agl unless in ATZ or controlled airspace with ATC permission
- for aerial work purposes unless with authority permission.

Microlight Aeroplane – ANO 118 (1)

A microlight is an aircraft not having a MSTOM over 390 kg, a wing loading at the MSTOM not over 25 kg/m^2, a maximum fuel capacity not over 50 litres and not designed to carry more than two people.

   Self-launching Motor Gliders (SLMG) will come into this category. JAA refers to these types of aeroplanes as Touring Motor Gliders (TMG), which

Table 21.1 Balloons, kites, airships and parascending parachutes

Type of aircraft	Clearance required from ground, etc.	When and where permitted	Height (agl)	Authority permission	Notes
Captive balloon	60 m from any vessel, vehicle or structure (VVS)*	Not within NOH of NATZ or airspace notified†	Not over 60 m	†Written permission may be obtained from CAA	*Unless person in charge of VVS permits. Must be securely moored and not unattended unless it automatically deflates if it breaks free
Tethered controllable balloon		Not within notified airspace or in NATZ*		*Permission may be obtained from ATC	
Controllable balloon in free flight		Not within notified airspace or in NATZ during NOH			Except during the day and VMC
Kite		Not in NOH of NATZ* and not at other times†	Not over 30m and †not over 60 m	*Written permission may be obtained from CAA	
Parascending, parachute	No launch by cable, winch, ground tow over 60 m (agl)	No launch within NOH of NATZ		Written permission may be obtained from CAA	
Glider	No launch by cable, winch, ground tow over 60 m (agl)			Written permission may be obtained from CAA	

Airship not over 3000 m³ moored securely and attended		At notified aerodrome and at other sites with permission*	*Written permission may be obtained from CAA
Airship not over 3000 m³ moored securely and attended	2 km from congested area unless permission given	At a notified aerodrome or in NATZ with permission*	*Written permission may be obtained from CAA
1000–2000 small balloons* simultaneously released		In NOH at NATZ	28 days' notice must be given to CAA	*Small balloons not over 2 m linear dimension at any time in flight
2000–10 000 small balloons* simultaneously released		Notified airspace or within NOH of NATZ	Written permission may be obtained from CAA	*Small balloons not over 2 m linear dimension at any time in flight
Over 10 000 small balloons* simultaneously released		At any site	Written permission may be obtained from CAA	*Small balloons not over 2 m linear dimension at any time in flight

are defined as: 'motor gliders having a C of A and an accompanying flight manual showing it to be capable of taking-off and climbing under its own power. It will have an integrally mounted non-retractable engine and a non-retractable propeller.' These aeroplanes will be certificated under JAR 22 – 'Sailplanes and Powered Sailplanes'.

Previously, the training on these aircraft was conducted by the British Gliding Association (BGA) but the CAA are now setting out approved training courses leading to a Private Pilot's Licence (SLMG).

ANO Interpretation – ANO 118

This is a useful article which gives a glossary of all the terms used in the ANO whose meaning may not be immediately clear. The definitions also apply to the AN(G)R and the R of A. It is a very extensive list and a selection is given here of only the most important ones not defined elsewhere in this book:

- *Aerial work*
  any purpose conducted for a valuable consideration which is not public transport; however, the following are deemed to be private flights:
  - aircraft competitions even though there may be some payments made
  - flight made for a registered charity carried out with the CAA's written permission
  - flights of four persons or less all sharing the cost
  - flights when the pilot reclaims the costs from the employer
  - flights in a jointly owned aircraft when the only payment is for the actual costs and is made by one or more of the owners.

  However, a flight in which 'valuable consideration' is involved for the carriage of passengers, which flight is for the purpose of dropping persons by parachute in accordance with CAA's permission is an aerial work flight. Similarly, positioning flights made prior to such a flight and the return of the aircraft to its base afterwards are also aerial work flights.

- *Authority or competent authority*
  in the UK this is the CAA and for other countries the body responsible under the law for promoting the safety of civil aviation.
- *Maximum Structural Take-off Mass (MSTOM)*
  is the maximum permitted mass of the aircraft under any conditions. It will be given in the Certificate of Airworthiness.

- *Night*
  in JAR-FCL, night is from the end of evening civil twilight to the beginning of morning twilight unless the local authority says otherwise. In the UK the times used are between 30 min after sunset until 30 min before sunrise at sea level. This amounts to practically the same times. In the UKAIP and similar documents HN (Heures Nuit) is defined as sunset to sunrise. HJ (Heures Jour) is sunrise to sunset and H24 all the hours of the day. These will be met when referring to aerodrome opening hours, radio facility operating times and active hours for restricted airspace.
- *Non-precision approach*
  is an approach not using ILS, MLS or PAR but some other radio aids. Associated with this will be the Minimum Descent Height (MDH) below which further descent must not be made without adequate visual reference for landing.
- *Precision approach*
  is an instrument approach using either Instrument Landing System (ILS) or the Microwave ILS (MLS) or a Precision Approach Radar (PAR) for azimuth and elevation guidance. Associated with this will be the Decision Height (DH) at which the pilot must initiate a missed approach procedure unless he or she has adequate visual reference for landing.
- *Public transport aircraft*
  is when the aircraft is flying with valuable consideration for the carriage of passengers or cargo. It includes the free carriage of the operator's staff or cargo.
- *Scheduled journey*
  is one of a series of journeys undertaken as part of a regular systematic service.
- *Valuable consideration*
  is any right, interest, profit or benefit, forbearance, detriment, loss or responsibility accruing, given, suffered or undertaken pursuant to an agreement which is of more than a nominal nature.

Test 1

(Numbers in brackets indicate the page where the answer will be found)

(1) Minimum weather conditions for a VFR flight below FL 100 in controlled airspace classes D and E are given in terms of minimum separation from cloud vertically and horizontally and a certain minimum visibility. These distances are respectively:
   - (a) 2000 ft   1500 m   5 km
   - (b) 500 ft   1800 m   8 km
   - (c) 1000 ft   1500 m   5 km
   - (d) 1000 ft   1800 m   5 km   (15)

(2) One of the following is required to obtain clearance for IFR flight at night in controlled airspace:
   - (a) a CPL
   - (b) an ATPL
   - (c) a night rating
   - (d) an instrument rating   (68)

(3) Within controlled airspace, air traffic clearance is required for:
   - (a) all flights under IFR
   - (b) only for IFR flights at night
   - (c) VFR flights at night
   - (d) VFR flights day and night   (65)

(4) Select the appropriate flight level if flying under IFR in the UK outside controlled airspace on a track of 270°(M)
   - (a) 55
   - (b) 60
   - (c) 65
   - (d) 70   (18)

(5) If on an airway and flying in a direction of 235°(M), your flight level will be:
   - (a) usually ODDs
   - (b) usually EVENs
   - (c) always ODDs
   - (d) always EVENs   (19)

(6) Flying VFR under ICAO rules, select a suitable FL if on a track of 180°(M):
   (a)   35
   (b)   40
   (c)   45
   (d)   50   (19)

(7) Flying IFR under ICAO rules, select a suitable FL if on a track of 175°(M):
   (a)   330
   (b)   340
   (c)   350
   (d)   360   (18)

(8) When flying an advisory route and wishing to use the advisory service, it is necessary to file a flight plan:
   (a)   only if flying under IFR
   (b)   only if flying within controlled airspace
   (c)   always
   (d)   only at night   (69)

(9) The ground signal consisting of a red square with one yellow diagonal indicates that the pilot should:
   (a)   take special care on the manoeuvring area
   (b)   not land at this aerodrome
   (c)   only land if it is an emergency situation
   (d)   keep to the hard surfaces when taxying   (52)

(10) A scheduled danger area is one that is active:
   (a)   always
   (b)   always at stated times
   (c)   when notified by NOTAMs
   (d)   from sunset to sunrise   (87)

(11) During an initial climb in uncontrolled airspace, the altimeter setting used should be:
   (a)   any desired value
   (b)   1013.2 mb
   (c)   the local QNH
   (d)   the local QFE   (24)

(12) Airways generally have a lower limit that is at least:
   (a)   3000 ft amsI
   (b)   3000 ft agI
   (c)   FL 30
   (d)   FL 40   (9)

(13) Controlled airspace is an area:
  (a)   only within an FIR in which an ATC service is provided
  (b)   only within a UIR in which an ATC service is provided
  (c)   within an FIR/UIR in which an ATC service is provided
  (d)   with a lower level which is above ground level   (8)

(14) To obtain a check on terrain clearance en-route, the altimeter setting to use will be:
  (a)   QNE
  (b)   local QNH
  (c)   local QFE
  (d)   regional QNH   (20)

(15) In Europe, flying VFR outside controlled airspace, below FL 300 and on a track of 056° (M) the cruising levels to be used will be:
  (a)   ODDs
  (b)   EVENs
  (c)   ODDs + 500 ft
  (d)   EVENs + 500 ft   (19)

(16) When flying in advisory airspace and using the advisory service, separation is provided from:
  (a)   all other traffic
  (b)   other IFR traffic
  (c)   other traffic using the advisory service
  (d)   other known traffic   (16)

(17) The controlling Authority for a TMA is generally:
  (a)   the ATC for the main airfield under the TMA
  (b)   the ATCC for the FIR
  (c)   the zone controller
  (d)   the ATC for the airfield directly beneath the TMA   (17)

(18) Pre-flight altimeter checks should be carried out:
  (a)   at the end of the runway while awaiting take-off clearance
  (b)   in the flight clearance office
  (c)   at the aerodrome reference point
  (d)   on the apron   (24)

(19) When an aircraft is approaching an aerodrome which is below a TMA, the altimeter setting to use, having passed through the transition altitude, is:
  (a)   regional QN
  (b)   regional QFE
  (c)   aerodrome QNH
  (d)   aerodrome QFE   (23)

(20) A flight plan must be filed when flying:
    (a)   an advisory route but not using the advisory service
    (b)   IFR in controlled airspace
    (c)   under a SVFR clearance
    (d)   more than 10 nm from the coast   (43)

(21) Following an AIRPROX incident a report should be made:
    (a)   immediately by radio
    (b)   immediately after landing
    (c)   within 7 days
    (d)   using the AFTN after landing   (92)

(22) Purple airways are given the following ICAO classification:
    (a)   A
    (b)   B
    (c)   C
    (d)   D   (76)

(23) En-route air navigation obstructions must be lit if they are above:
    (a)   150 m amsI
    (b)   150 m agl
    (c)   100 m amsI
    (d)   l00 m agI   (90)

(24) A continuous red beam directed at an aircraft from the control tower means:
    (a)   owing to aerodrome being unserviceable, land elsewhere
    (b)   do not land, aerodrome closed
    (c)   do not land, give way to other aircraft and continue circling
    (d)   move clear of the landing area   (55)

(25) If a pilot lands at an aerodrome other than that given in the flight plan he must inform the original destination:
    (a)   within 30 min of his original ETA, of his landing elsewhere
    (b)   within 30 min of his landing elsewhere
    (c)   within 45 min of his original ETA of his landing elsewhere
    (d)   within 45 min of his landing elsewhere   (44)

(26) In Europe, Upper Airspace extends upwards from:
    (a)   FL 195
    (b)   FL 245
    (c)   FL 255
    (d)   FL varies but is always shown on the front of en-route charts   (8)

(27)   To fly IFR in controlled airspace, unless on a notified route or taking-off or landing, the minimum height to be flown is:
   (a)   150 m (500 ft) above all obstacles within 5 nm
   (b)   150 m (500 ft) above all obstacles within 10 nm
   (c)   300 m (1000 ft) above all obstacles within 5 nm
   (d)   300 m (1000 ft) above all obstacles within 10 nm   (18)

(28)   When flying over the sea an aircraft should not fly closer to a vessel than:
   (a)   600 m (2000 ft)
   (b)   450 m (1500 ft)
   (c)   300 m (1000 ft)
   (d)   l00m (500 ft)   (94)

(29)   Smoking is prohibited in an aircraft:
   (a)   whenever the commander thinks it necessary to forbid it
   (b)   during take-offs and landings
   (c)   when flying through turbulent air
   (d)   for the flight crew at any time   (155)

(30)   When carrying out a circuit at an aerodrome, the turns are normally:
   (a)   right-handed
   (b)   left-handed
   (c)   right- or left-handed depending on the runway in use
   (d)   right-handed unless a green flag is flying   (56)

(31)   When flying at night you see another aircraft moving at 90° and crossing your track from left to right. You would expect to see:
   (a)   a red light
   (b)   a green light
   (c)   a white light
   (d)   an anti-collision light only   (176)

(32)   When two aircraft are approaching head-on, the rules of the air require the following action:
   (a)   both aircraft alter heading to the right
   (b)   both aircraft alter heading to the left
   (c)   the larger aircraft alters heading right
   (d)   the smaller aircraft alters heading right   (172)

(33)   For an aircraft on an international private flight the following documents must be carried:
   (a)   Certificate of Airworthiness only
   (b)   Certificate of Registration only
   (c)   crew licences only
   (d)   all of these   (163)

(34) When navigating visually using a line feature such as a railway, the pilot should keep the feature:
  (a) always on the left of the aircraft
  (b) on the left of the aircraft unless ATC requires otherwise
  (c) always on the right of the aircraft
  (d) directly beneath the aircraft   (174)

(35) Details of a temporary danger area being introduced will always be given in:
  (a) NOTAMs
  (b) Yellow Aeronautical Information Circulars (AIC)
  (c) Mauve AIC
  (d) Aeronautical Information Publication   (88)

(36) In flight at night you observe a white flashing light ahead which is getting closer, you could be:
  (a) seeing an aircraft anti-collision light
  (b) overtaking another aircraft
  (c) approaching an aerodrome beacon
  (d) any of the above   (37 and 175)

(37) Customs aerodromes are:
  (a) open for 24 hours
  (b) usually open from 0600 to 2000 UTC
  (c) open to meet the regular traffic needs
  (d) have customs service only when requested   (118)

(38) A passenger who appears to be under the influence of alcohol so that the operator believes him to be a danger to the aircraft or its occupants:
  (a) shall not permit him/her to board
  (b) shall be permitted on board only if accompanied by a sober companion
  (c) shall be required to take a blood alcohol level test
  (d) shall be required to be breathalysed   (143)

(39) An aircraft shall not fly over an open air assembly of a lot of people (say 1000) within
  (a) 300 m (1000 ft)
  (b) 450m (1500 ft)
  (c) 600 m (2000 ft)
  (d) 900 m (3000 ft)   (93)

(40) An authority representative responsible for certification, licensing or inspection may be admitted to the flight deck at any time:
  (a) during flight
  (b) only if the commander is sure that safety will not be affected

    (c)   only during cruising flight

    (d)   except for take-off and landing   (144)

(41) The letter 'C' displayed in black against a yellow background indi-cates the place where:

    (a)   flight plans should be filed or booking-out carried out

    (b)   air traffic control is situated

    (c)   pilots of visiting aircraft should report

    (d)   aeronautical information service is available   (53)

(42) When approaching an aerodrome under VFR where there is no ATC operational, the pilot should keep:

    (a)   clear of cloud and not descend to below 150 m (500 ft) above aerodrome level

    (b)   clear of cloud and not descend to below 300 m (1000 ft) above aerodrome level

    (c)   300 m (1000 ft) below cloud

    (d)   150 m (500 ft) below cloud   (62)

(43) When one taxying aircraft wishes to overtake another, the overtak-ing aircraft:

    (a)   should alter heading to the right

    (b)   should alter heading to the left

    (c)   should alter heading left or right

    (d)   should refrain from overtaking   (173)

(44) An authorised person may require the production of the Certificate of Airworthiness. The responsibility lies with the:

    (a)   operator, who should produce it within 5 days

    (b)   commander, who should produce it in 5 days

    (c)   commander, who should produce it in a reasonable time

    (d)   operator, who should produce it in a reasonable time   (162)

(45) A pilot's flying log book should be produced, when requested by an authorised person, within:

    (a)   a reasonable time

    (b)   within 5 days

    (c)   within 7 days

    (d)   within 14 days   (169)

(46) The marshalling signal 'arms extended, palms facing inwards, arms swung from the extended position inwards' means:

    (a)   this bay

    (b)   stop

    (c)   chocks inserted

    (d)   chocks removed   (179)

(47) An aeroplane is landing at an aerodrome without runways and without air traffic control. There is another aircraft ahead. The pilot, with reference to the aircraft in front, should leave it clear on his/her:
(a) left and subsequently turn left
(b) left and subsequently turn right
(c) right and subsequently turn left
(d) right and subsequently turn right   (62)

(48) Without special permission, the last airport to take-off from for a flight to outside the European Union will be:
(a) a sanitary aerodrome
(b) a customs aerodrome
(c) a licensed aerodrome
(d) any aerodrome   (119)

(49) The aerodrome beacons at civil aerodromes now usually show a:
(a) white flashing strobe light
(b) two-letter identification code in red
(c) two-letter identification code in green
(d) two-letter identification code in white   (37)

(50) When a special meteorological forecast is required for a flight in excess of 500 nm, the following notice should be given:
(a) 4 hours
(0) 3 hours
(c) 2 hours
(d) 1 hour   (97)

Answers to Test 1

Question	Answer	Question	Answer
1	c	26	d
2	d	27	c
3	a	28	d
4	c	29	a
5	b	30	b
6	c	31	b
7	a	32	a
8	c	33	d
9	a	34	b
10	b	35	a
11	a	36	d
12	b	37	c
13	c	38	a
14	d	39	d
15	c	40	b
16	c	41	c
17	b	42	a
18	d	43	b
19	c	44	c
20	b	45	a
21	a	46	c
22	a	47	a
23	b	48	b
24	c	49	a
25	a	50	a

Test 2

(Numbers in brackets indicate the page where the answer will be found)

(1) The elevation is the altitude of the:
    (a)   highest point on the landing area
    (b)   lowest point on the manoeuvring area
    (c)   highest point on the runway
    (d)   highest point on the aerodrome   (21)

(2) The minimum crew that must be carried on a public transport flight is stated in:
    (a)   JAR-OPS
    (b)   the Air Navigation Order
    (c)   the aeroplane flight manual
    (d)   the operations manual   (131)

(3) Runway visual range is reported when the its value falls below:
    (a)   1000 m
    (b)   1200 m
    (c)   1500 m
    (d)   1800 m   (37)

(4) The entries in a pilot's personal flying log book must include:
    (a)   name and address of the pilot and his/her employer (if any)
    (b)   full details of every flight made
    (c)   details of any test carried out in a simulator
    (d)   all of these   (167)

(5) The take-off distance available at an aerodrome consists of:
    (a)   the take-off run available plus the stopway
    (b)   the take-off run available plus the clearway
    (c)   the take-off run available plus the stopway plus the clearway
    (d)   the take-off run available   (32)

(6) An aircraft flying to or from abroad may cross the UK coastline:
    (a)   only at a compulsory reporting point
    (b)   only at a point designated in the flight plan
    (c)   at any point providing it complies with ATS routes
    (d)   at any point which is not in a prohibited area   (116)

(7) At an aerodrome where there is only a Aerodrome Flight Information Service Officer (AFISO), for landing the pilot will:
   (a) use the direction indicated by the ground signals or, in their absence, land into wind unless good aviation practice demands otherwise
   (b) follow the advice of the AFISO
   (c) obey the instructions in his flight manual
   (d) obey the instructions in his operations manual (61)

(8) EAT is the time that an aircraft:
   (a) will be expected to join the holding stack
   (b) will leave the last en route holding point before the destination
   (c) will be expected to leave the holding stack and approach the aerodrome
   (d) is expected to be landing (57)

(9) For first-aid purposes, a supply of pure oxygen must be available in the passenger cabin of a pressurised aircraft if it is to be operated over a pressure altitude of over:
   (a) 10 000 ft
   (b) 13 000 ft
   (c) 20 000 ft
   (d) 25 000 ft (156)

(10) The maximum permitted flight time for flight crew is:
   (a) 100 hours in the 27 days prior to the present flight
   (b) 69 hours during the 27 days prior to the flight
   (c) 1000 hours in the year up to the end of the previous month to the present flight
   (d) 1200 hours in the year up to the end of the previous month to the flight (146)

(11) The Aerodrome Operating Minima (AOM) for an aircraft landing are:
   (a) decision height and runway visual range when using a precision approach aid
   (b) decision height and runway visual range when using a non-precision approach aid
   (c) minimum descent height and runway visual range when using a precision approach aid
   (d) minimum descent height and visibility when using a non-precision approach aid (154)

(12) The weather information passed by approach control to arriving aircraft is given in the following order:
  (a)   cloud ceiling and RVR
  (b)   cloud, weather, surface wind and visibility
  (c)   runaway in use, surface wind, visibility, weather, altimeter settings, weather warnings, RVR
  (d)   just the information requested by the pilot   (58)

(13) Transition altitude is the altitude in the vicinity of an aerodrome at which the pilot:
  (a)   on the descent changes the altimeter setting to 1013.2 mb
  (b)   on climbing away from the aerodrome changes the altimeter setting to 1013.2 mb
  (c)   on climbing away from the aerodrome changes the altimeter setting to QNH
  (d)   on climbing away from the aerodrome changes the altimeter setting to QFE   (21)

(14) QFE threshold is passed to the pilot if the threshold of a precision approach runway is over:
  (a)   7 ft above the airfield elevation
  (b)   7 ft below the airfield elevation
  (c)   15 ft above the airfield elevation
  (d)   15 ft below the airfield elevation   (20)

(15) An operator is responsible for ensuring that correct ditching equipment is carried in aeroplanes when flying over water more than 2 hours or 400 nm, whichever is the smaller, from where an emergency landing can be made, if the passenger seating capacity is over:
  (a)   30
  (b)   40
  (c)   50
  (d)   60   (149)

(16) A British Airways flight taking paying traffic from the UK, overflying the Irish Republic and landing in the USA will be using the following freedoms:
  (a)   1 and 4
  (b)   2 and 4
  (c)   2 and 3
  (d)   1 and 3   (73)

(17)   An AIRAC amendment:
    (a)   amends the En-Route section of the AIP
    (b)   revises a previous Aeronautical Information Circular (AIC)
    (c)   is issued under the Regulated System for Air Information
    (d)   corrects a previous NOTAM   (25)

(18)   In an aerodrome circuit, the call 'Final' is made when:
    (a)   abeam upwind end of the runway
    (b)   on completion of the turn on to the base leg
    (c)   when making a straight-in approach at 6 nm out
    (d)   after completion of the turn on to the final approach and not
        more than 4 nm from the runway   (56)

(19)   An aircraft may be allowed to land on a runway before the preceding aircraft has cleared it providing:
    (a)   only that it is daylight
    (b)   only that it is daylight and the preceding aircraft is clearly visible throughout the landing period
    (c)   only that the runway is long enough
    (d)   all the above   (56)

(20)   If, when waiting at the end of the runway prior to take-off at night, ATC advise that a navigation light has failed, the pilot:
    (a)   may take off, if ATC agree
    (b)   may take off, if the anti-collision light is working
    (c)   may not depart
    (d)   may depart as long as the light is replaced at the
        next landing   (177)

(21)   A pilot who suffers an injury making him/her incapable of undertaking the duties for which the licence has been issued should inform the Authority:
    (a)   when 20 days have elapsed
    (b)   when advised to do so by a medical doctor
    (c)   within 20 days
    (d)   as soon as possible   (146)

(22)   The load sheet is signed before a public transport flight by the:
    (a)   load supervisor only
    (b)   flight commander only
    (c)   load supervisor and the flight commander
    (d)   operator, the load supervisor and the flight commander (150)

(23)   The separation distance for traffic when only secondary surveillance radar is available is:
    (a)   8 nm

(b)   5 nm

(c)   3 nm

(d)   2 nm   (40)

(24)   The en-route vertical clearance from obstacles used by the Radar
Services, if aircraft flying under IFR are not on airways or advisory
routes, is:
(a)   1000 ft over the highest obstacle within 10 nm
(b)   1000 ft over the highest obstacle within 30 nm
(c)   1000 ft over the highest obstacle within 15 nm
(d)   1500 ft over the highest obstacle within 15 nm   (81)

(25)   If radio navigation equipment fails, the correct immediate action is
to:
(a)   leave controlled airspace immediately
(b)   inform ATC, giving altitude and approximate position
(c)   ensure it is made serviceable at the next landing
(d)   leave or avoid entering controlled airspace and continue flight
in VMC   (48)

(26)   If an aircraft's secondary surveillance radar fails before take-off:
(a)   go to the nearest aerodrome where it can be repaired
(b)   inform ATC who may give permission to proceed subject to
certain conditions
(c)   insert 'N' in item 10 of the Flight Plan
(d)   do all of these   (49)

(27)   Above 30 000 ft the vertical flight separation between aircraft on
reciprocal tracks will be:
(a)   4000 ft
(b)   2000 ft
(c)   1000 ft
(d)   1000 or 2000 ft, depending on whether the aircraft is in
Reduced Vertical Separation Minima (RVSM) airspace or
not   (19)

(28)   A SIGMET message concerns:
(a)   weather phenomena which may affect the safety of aircraft
operations and is issued by a Meteorological Watch Office
(MWO)
(b)   a routine report of significant weather issued by the local
Meteorological Office
(c)   weather phenomena which may affect the safety of aircraft
operations and is issued by a local Meteorological Office
(d)   severe weather encountered by the pilot of an aircraft   (99)

(29) Clearway is the area from:
   (a)   the end of the runway suitable for the ground run of an aircraft that is decelerating after an aborted take-off
   (b)   the end of the stopway suitable for the ground run of an aircraft that is decelerating after an aborted take-off
   (c)   the end of the take-off run available which is selected as suitable for an aircraft to make its initial climb to a specified screen height
   (d)   the end of the stopway which is selected as suitable for an aircraft to make its initial climb to a specified screen height (32)

(30) HN and 'night' according to JAR-FCL have different meanings. These are from:

HN		'Night'
(a)	sunset to sunrise	end of evening Civil Twilight (CT) to beginning of morning CT
(b)	sunset to sunrise	30 min after sunset to 30 min before sunrise
(c)	1800 to 0600 LT	30 min after sunset to 30 min before sunrise
(d)	1800 to 0600 LT	end of evening Civil Twilight (CT) to beginning of morning CT

(187)

(31) According to JAR-OPS, standard masses of passengers may be used for preparing the load sheet only when the aircraft has seats for over the following number of passengers:
   (a)   10
   (b)   15
   (c)   20
   (d)   30   (150)

(32) The ground signal displayed by survivors requiring medical assistance is:
   (a)   V
   (b)   X
   (c)   N
   (d)   Y   (113)

(33) The recommended times for operating the old-style cranked dinghy radio are:
   (a)   on the hour for 3 min
   (b)   on the hour and half-hour for 3 min
   (c)   continuously whenever search aircraft are heard   (114)

(d)   quarter past and quarter to the hour for 3 min

(34) The responsibility for keeping the operations manual amended rest with the:
(a)   operator
(b)   the aircraft commander
(c)   the Authority
(d)   ground operations manager   (148)

(35) According to JAR-OPS, an operator may use his own values for the standard masses of passengers provided that:
(a)   he has given notice to the Authority that he intends to do so
(b)   the commander of the aircraft is prepared to accept the values being used
(c)   he has carried out an elaborate sampling procedure and submitted the proposed values to the Authority for approval
(d)   the values to use have been indicated in the aeroplane flight manual   (150)

(36) If you are at the correct height on approach to the runway equipped with PAPI you should see:
(a)   one white light and two reds
(b)   one white light and one red
(c)   three white lights and one red
(d)   two white lights and two red   (36)

(37) If a controller responding to a distress message asks the pilot to use the speechless code and the pilot responds by pressing his transmit button for three short transmissions this signifies:
(a)   yes
(b)   no
(c)   say again
(d)   request homing   (115)

(38) A public transport aircraft shall not take off unless the following minima for the departure airfield are satisfactory:
(a)   cloud base and visibility
(b)   cloud base and RVR
(c)   cloud ceiling and RVR
(d)   minimum descent height and RVR   (154)

(39)   An aircraft commander should transmit special aircraft observations whenever he encounters
   (a)   severe turbulence or severe icing
   (b)   moderate turbulence, hail or cumulonimbus clouds during transonic or supersonic flight
   (c)   other conditions that the aircraft commander considers may affect the safety of other aircraft
   (d)   all of these   (101)

(40)   Under UK law the following accident is 'reportable' to the Authority:
   (a)   a passenger climbing up the stairs attached to the aircraft slips is seriously injured although the steward at the top of the stairs tried to warn him of danger
   (b)   an aircraft on landing bursts a tyre
   (c)   a passenger on the way out to the aircraft falls and is seriously injured
   (d)   an attempted hijack is thwarted and the hijacker is seriously injured   (166)

(41)   A public transport aircraft flying on ATS routes must carry B-RNAV equipment on:
   (a)   every flight
   (b)   any flight going outside the European Civil Aviation Conference (ECAC) area
   (c)   any flight inside the ECAC area over a specified level (usually at least FL 100)
   (d)   any flight inside the ECAC area   (47)

(42)   Inside controlled airspace, the flight levels allocated by ATC will be:
   (a)   ODDs or EVENs
   (b)   ODDS if flying easterly
   (c)   EVENs if flying easterly
   (d)   EVENs if flying westerly   (19)

(43)   On a precision approach runway, touch-down zone markings are placed:
   (a)   150 m apart
   (b)   300 m apart
   (c)   450 m apart
   (d)   600 m apart   (35)

(44) A class 'B' bearing from a VHF/DF station should be accurate within:
- (a)   ±1°
- (b)   ±2°
- (c)   ±5°
- (d)   ±10°   (104)

(45) A flight plan should be filed by the following time before taxi clearance is required:
- (a)   never less than 60 min
- (b)   normally 60 min but never less than 30 min
- (c)   never less than 2 hours
- (d)   never less than 3 hours   (33)

(46) An area Flight Information Service (FIS) controller may provide the following service to aircraft in the FIR:
- (a)   headings to steer to avoid other traffic
- (b)   a service for crossing controlled airspace
- (c)   a complete picture of all proximity hazards
- (d)   none of these   (73)

(47) A white flashing light from the ATC to a taxiing aircraft means:
- (a)   give way to other aircraft on the taxiway
- (b)   return to the starting point on the aerodrome
- (c)   give way to aircraft about to land
- (d)   await further instructions   (55)

(48) One of the conditions that must be complied with when carrying our simulated instrument flight:
- (a)   an observer is carried in the second pilot's seat
- (b)   the aircraft has dual controls
- (c)   an observer is carried in the second pilot's seat and, if his field of vision is not adequate, a third person is carried
- (d)   screens must be fitted to restrict the pilot's vision   (174)

(49) Anti-collision lights fitted to aircraft should be omni-directional flashing lights fitted as follows:

	To aeroplanes	To helicopters
(a)	only white	only red
(b)	only red	only red
(c)	red or white	only red
(d)	red or white	only white

(177)

(50) An aircraft lands at an aerodrome and requires the replacement of a part included in the Certificate of Airworthiness. A Certificate of Release cannot be obtained and the aircraft has to fly to another aerodrome to obtain it. The commander of the aircraft should notify the Authority giving full details:

(a) within a reasonable time

(b) as soon as the aircraft is back in its country of registration

(c) within 10 days

(d) immediately after landing at the other aerodrome   (128)

Answers to Test 2

Question	Answer	Question	Answer
1	a	26	d
2	d	27	d
3	c	28	a
4	d	29	c
5	b	30	a
6	c	31	c
7	a	32	b
8	c	33	d
9	d	34	a
10	a	35	c
11	a	36	d
12	c	37	c
13	b	38	c
14	b	39	d
15	a	40	a
16	d	41	c
17	c	42	a
18	d	43	a
19	d	44	c
20	c	45	b
21	d	46	b
22	c	47	b
23	a	48	b
24	b	49	c
25	b	50	c

Test 3

The aviation law examination always features a number of questions based on aircraft lights and the anti-collision rules. This short test gives examples of these. It will be helpful to refer to the rhyme featured in Chapter 20.

Questions 1–8

Answer these question by choosing the correct response from the following four alternatives:

    A   there is no risk of collision
    B   there is a risk of collision so turn left
    C   there is a risk of collision so turn right
    D   there is a risk of collision so maintain course and speed but be prepared to take avoiding action if the other aircraft fails to give way

Question	Lights seen in addition to anti-collision light	Relative bearing of light(s) when first seen	Change of relative bearing, if any
(1)	Red	040°	none apparent
(2)	Red and green	350°	none apparent
(3)	Green	340°	none apparent
(4)	Green	040°	none apparent
(5)	white flashing in alternation with white	005°	none apparent
(6)	white flashing alternating with red and green lights	350°	none apparent
(7)	white flashing alternating with green	310°	decreasing
(8)	white flashing alternating with red and green lights	030°	increasing

(9) You are flying on a heading of 290° (T) when you see a green light on a relative bearing of 320°. Between what headings could the other aircraft be flying?:

(a)   150° and 260°
(b)   320° and 070°
(c)   070° and 125°
(d)   015° and 070°

(10) You are flying on a heading of 180° (T) when you see a white light on a relative bearing of 350°. Between what headings could the other aircraft be flying?:

(a)   240° through north to 100°
(b)   240° through south to 100°
(c)   340° through north to 030°
(d)   340° through south to 030°

(11) Flying at night you see the red and green lights of another aircraft at a range of about 4 nm and at about your altitude on a relative hearing of 085°. Is there a risk of collision and what action would you take?

	Risk of collision	Action to take
(a)	Yes	Maintain heading
(b)	Yes	Turn right
(c)	Yes	Turn left
(d)	No	None

Answers to Test 3

(1) c   Referring to rhyme, Red on Green requires action so turn right.

(2) c   No change of bearing so aircraft is approaching head on and action required, so turn right.

(3) d   Referring to rhyme, Green on Red requires action by the other aircraft.

(4) a   A changing bearing indicates no risk of collision.

(5) c   Approaching the rear of another aircraft with no change of bearing so there is a collision risk and so turn right.

(6) c   Approaching head-on to the other aircraft with no change of bearing so there is a collision risk and so turn right.

(7) a   A changing bearing indicates no risk of collision

(8) a   A changing bearing indicates no risk of collision

(9) b   The other aircraft is converging. 290° + 320° = 610° – 360° = 250° – this is the true bearing of the other aircraft from you. Your bearing from the other aircraft will be 250° – 180° = 070°. The green light is visible through 110° and so the other aircraft can be on 070° or 070° – 110° or 430° – 110° = 320°. The other aircraft is heading between 320° through north to 070°.

(10) b   You are approaching the other aircraft from the rear. Its white light will be visible through a 140° arc and so the other aircraft can be flying the same heading as you or your heading ±70° (140° ÷ 2 = 70°). 170° ± 70° = 100° to 240° through south.

(11) d   You are looking at the other aircraft from a head-on position but at a relative bearing of 085° and so you obviously going to cross the other aircraft's track and there will be no collision risk.

Test 4

The Authority requires candidates to have knowledge of important Aeronautical Information Circulars (AIC). This test is designed to direct your attention to important circulars. Because the AICs are being constantly reissued, the numbers will be changing and so each question is annotated with the colour and title of the relevant AIC. Using the index, which is issued at regular intervals, there should be no difficulty in tracking down the circular required to answer the questions.

(1) On some aircraft, icing of engine pressure probes can cause an over-reading of engine instruments. To minimise this possibility, in the absence of specific information, it is recommended that this is likely to occur when:

	Outside air temperature (°C)	RVR
(a)	less than +5	less than 1000 m
(b)	less than +5	less than 1500 m
(c)	less than +10	less than 1500 m
(d)	less than +10	less than 1000 m

(Pink AIC – Frost, Ice and Snow on Aircraft)

(2) At present approved Global Positioning System (GPS) equipment may be used for:

	Basic area navigation (B-RNAV) Routes	Instrument approaches
(a)	No	No
(b)	No	Yes
(c)	Yes	No
(d)	Yes	Yes

(Pink AIC – UK Policy for the use of United States Navstar GPS)

(3) An aircraft, in wake turbulence category medium, is taking off from the same place on the runway as an aircraft in the heavy category which has just taken off. The ATC will allow a minimum space in the airborne times of:
(a) 1 min
(b) 2 min
(c) 3 min
(d) 4 min
(Pink AIC – Wake Turbulence)

(4) The objective of the occurrence reporting system is to ensure that:
  (a)   the Authority is made aware of hazardous or potentially hazardous incidents
  (b)   the information of the occurrences is disseminated
  (c)   the Authority can evaluate the safety implications of the occurrences and take any necessary action to avoid them happening again
  (d)   all of these
  (Pink AIC – Occurrence Reporting)

(5) When transmitting an URGENCY message, a student pilot should use the prefix:
  (a)   PAN PAN and include TYRO in the message
  (b)   PAN PAN and include STUDENT in the message
  (c)   MAYDAY and include TYRO in the message
  (d)   MAYDAY and include STUDENT in the message
  (Pink AIC – Use of VHF International Emergency Service)

(6) Departing from an ATC clearance in response to a Resolution Advisory (RA) from an Airborne Collision Avoidance System (ACAS) is permitted but the Commander should:
  (a)   obtain ATC permission before acting on the RA
  (b)   advise ATC immediately that he is acting on the RA
  (c)   give written particulars to the Authority within 10 days of the incident
  (d)   both (b) and (c) must be done
  (Pink AIC – ACAS – Legal Aspects and Interface with ATC)

(7) For JAR-OPS performance purposes, runways will be considered as *contaminated* if they are reported as being:
  (a)   damp
  (b)   wet
  (c)   with water patches
  (d)   any of these
  (Pink AIC – Risks and Factors Associated with Operations on Runways Affected by Snow, Slush or Water)

(8) Portable telephones may be carried on board an aircraft during flight, provided that they are:
  (a)   switched off prior to engine start-up until engine shut-down
  (b)   left on stand-by prior to engine start-up until engine shut-down
  (c)   switched off during the initial climb and on the final descent
  (d)   on stand-by during the initial climb and on the final descent
  (Pink AIC – Use of Portable Telephones in Aircraft)

(9)   A passenger is seriously ill and the pilot is advised by a doctor that he requires urgent medical attention. The pilot wishes to land a soon as possible and he should inform the ATC and:

(a)   prefix the message with PAN PAN
(b)   prefix the message with MAYDAY
(c)   prefix the message with MEDICAL URGENCY
(d)   explain to ATC the circumstances of the situation and request priority in the landing sequence

(Pink AIC – Medical Emergencies)

(10)   A drug recommended for the treatment of malaria if contracted by a member of aircrew is:

(a)   Halfan
(b)   Lariam
(c)   Paludrine
(d)   Avlocolor

(Pink AIC – Malaria)

(11)   The effect of an uphill runway gradient of 2% on a light aircraft will be to:

(a)   increase the take-off distance required by 20%
(b)   decrease the take-off distance required by 20%
(c)   increase the take-off distance required by 10%
(d)   decrease the take-off distance required by 10%

(Pink AIC – Take-off, Climb and Landing Performance of Light Aeroplanes)

(12)   A safety review of aircraft level violations showed that by far and away the commonest cause was:

(a)   exceeding the limits laid down in the Standard Instrument Departure Procedures (SID)
(b)   non-compliance with correctly read back clearances
(c)   inadequate flight planning
(d)   excessive workload

(Pink AIC – Level Violations)

(13)   The time intervals allowed to elapse before aircrew fly, after having had an anaesthetic should be:

	After local anaesthetic	After general anaesthetic
(a)	8 hours	12 hours
(b)	12 hours	24 hours
(c)	24 hours	36 hours
(d)	24 hours	48 hour

(Pink AIC – Medication, Alcohol and Flying)

(14) If you are flying in a thunderstorm, which of the following must you *not* try to do?
    (a)   climb to try to get on top
    (b)   control the aircraft regardless of all else
    (c)   maintain the original heading
    (d)   not attempt to maintain altitude gained or lost unless absolutely necessary
(Pink AIC – The Effect of Thunderstorms)

(15) A disused aerodrome where the runway has been inspected within the last 6 months and was then considered fit for emergency use, will have the following markings on the end of the runway:
    (a)   white cross
    (b)   white cross underlined by a white bar
    (c)   EO (emergency only)
    (d)   there is no indication in general use
(Pink AIC – Hazards in Using Disused Aerodromes)

(16) Which of the following requires advice from an Authorised Medical Examiner (AME) before returning to flying duty?
    (a)   any surgical operation
    (b)   any regular use of medication
    (c)   kidney stone treatment
    (d)   any of these
(Pink AIC – Modern Medical Practice and Flight Safety)

(17) Which of the following aircraft equipment is most susceptible to interference from passengers' portable electronic equipment?
    (a)   Automatic Direction Finders (ADF)
    (b)   Instrument Landing System (ILS)
    (c)   Distance Measuring Equipment (DME)
    (d)   Flight Management Systems (FMS)
(Pink AIC – Use of Portable Electronic Games, Calculators, etc. in aircraft)

(18) The Designated Operational Coverage (DOC) for the following radio aids is specified as:

	VOR	ADF
(a)	range	range and altitude
(b)	altitude	range
(c)	range and altitude	range
(d)	range and altitude	range and altitude

(Pink AIC – Radio Navigation Aids – Designated Operational Coverage)

(19) Excess height at the threshold for a large transport aeroplane can result in a greater landing run. The additional runway length required for every 10 ft of excess height is:
(a)   100 ft
(b)   200 ft
(c)   250 ft
(d)   300 ft
(Pink AIC – Landing Performance of Large Transport Aeroplanes)

(20) Aircrew should not fly within the following number of hours after donating bone marrow, which involved a general anaesthetic, or donating blood:

	Blood	Bone marrow
(a)	24	48
(b)	24	36
(c)	12	48
(d)	12	36

(Pink AIC – Blood, Plasma and Bone Marrow Donation)

(21) if ATC give the message 'No delay expected', this indicates that the aircraft is not expected to be in the holding pattern for more than the following number of minutes:
(a)   10
(b)   15
(c)   20
(d)   25
(Pink AIC – Aeroplanes Inbound to the UK with Fuel Reserves Approaching Minimum)

Answers to Test 4

Question	Answer	Question	Answer
1	d	11	c
2	c	12	b
3	b	13	d
4	d	14	a
5	a	15	b
6	d	16	d
7	c	17	a
8	a	18	c
9	b	19	b
10	a	20	a
		21	c

Index